Biotechnology Demystified

Jaxon Garrison

SYRAWOOD
PUBLISHING HOUSE

New York

Published by Syrawood Publishing House,
750 Third Avenue, 9th Floor,
New York, NY 10017, USA
www.syrawoodpublishinghouse.com

Biotechnology Demystified
Jaxon Garrison

International Standard Book Number: 978-1-64740-087-3 (Hardback)

Cataloging-in-Publication Data

Biotechnology demystified / Jaxon Garrison.
    p. cm.
Includes bibliographical references and index.
ISBN 978-1-64740-087-3
1. Biotechnology. 2. Genetic engineering. 3. Microbiology. I. Garrison, Jaxon.
TP248.2 .B56 2022
660.6--dc23

Table of Contents

Permissions

Index

Preface

This book is a culmination of my many years of practice in this field. I attribute the success of this book to my support group. I would like to thank my parents who have showered me with unconditional love and support and my peers and professors for their constant guidance.

Biotechnology is a field of biology that involves living systems and organisms for the production and development of products. It is an interdisciplinary field which incorporates principles from molecular biology, biomedical engineering and biomanufacturing. Biotechnology applies concepts from numerous fields such as agriculture, food production, medicine, genomics and immunology. Some of the important sub-disciplines within this field are blue biotechnology, bioinformatics, green biotechnology and red biotechnology. The applications of biotechnology are categorized into four areas, namely, health care, crop production, industrial uses of crops, and production of other products such as biodegradable plastics and biofuels. This textbook is a compilation of chapters that discuss the most vital concepts in the field of biotechnology. The various sub-fields of biotechnology along with technological progress that have future implications are glanced at in it. Through this book, we attempt to further enlighten the readers about the new concepts in this field.

The details of chapters are provided below for a progressive learning:

Chapter – What is Biotechnology?

Biotechnology is a discipline that comprises of a number of procedures for the modification of living organisms as per human purposes. Bioinformatics, green biotechnology, blue biotechnology, red biotechnology are the various branches that fall under this domain. This is an introductory chapter which will briefly introduce biotechnology as well its branches.

Chapter – Concepts in Biotechnology

The fundamental concepts of biotechnology are related to cell, gene, proteins, DNA, RNA, fermentation and the human genome project. The topics will help in gaining a better perspective about these concepts in biotechnology.

Chapter – Fundamental Techniques in Biotechnology

Various techniques and methods are used in biotechnology. A few of them are cell culture, tissue culture, cloning, selective breeding, bioremediation and biosynthesis. The chapter closely examines these fundamental techniques in biotechnology to provide an extensive understanding of the subject.

Chapter – Applications of Biotechnology

Biotechnology is applied in various disciplines such as genetic engineering, biomimetics and biomanufacturing. It also plays an important role in the functioning of a biobased economy. The diverse applications of biotechnology in these areas have been thoroughly discussed in this chapter.

Chapter – Fields Related to Biotechnology

Environmental biotechnology, agricultural biotechnology, industrial biotechnology, bioinformatics, biomedical engineering, nanobiotechnology, biomedicine and molecular biology are some of the fields which are closely associated with biotechnology. All these diverse fields of biotechnology have been carefully analyzed in this chapter.

Jaxon Garrison

1

What is Biotechnology?

Biotechnology is a discipline that comprises of a number of procedures for the modification of living organisms as per human purposes. Bioinformatics, green biotechnology, blue biotechnology, red biotechnology are the various branches that fall under this domain. This is an introductory chapter which will briefly introduce biotechnology as well its branches.

Biotechnology refers to the use of biology to solve problems and make useful products. The most prominent area of biotechnology is the production of therapeutic proteins and other drugs through genetic engineering.

People have been harnessing biological processes to improve their quality of life for some 10,000 years, beginning with the first agricultural communities. Approximately 6,000 years ago, humans began to tap the biological processes of microorganisms in order to make bread, alcoholic beverages, and cheese and to preserve dairy products. But such processes are not what is meant today by biotechnology, a term first widely applied to the molecular and cellular technologies that began to emerge in the 1960s and '70s. A fledgling "biotech" industry began to coalesce in the mid- to late 1970s, led by Genentech, a pharmaceutical company established in 1976 by Robert A. Swanson and Herbert W. Boyer to commercialize the recombinant DNA technology pioneered by Boyer, Paul Berg, and Stanley N. Cohen. Early companies such as Genentech, Amgen, Biogen, Cetus, and Genex began by manufacturing genetically engineered substances primarily for medical and environmental uses.

For more than a decade, the biotechnology industry was dominated by recombinant DNA technology, or genetic engineering. This technique consists of splicing the gene for a useful protein (often a human protein) into production cells—such as yeast, bacteria, or mammalian cells in culture—which then begin to produce the protein in volume. In the process of splicing a gene into a production cell, a new organism is created. At first, biotechnology investors and researchers were uncertain about whether the courts would permit them to acquire patents on organisms; after all, patents were not allowed on new organisms that happened to be discovered and identified in nature. But, in 1980, the U.S. Supreme Court, in the case of Diamond v. Chakrabarty, resolved the matter by ruling that "a live human-made microorganism is patentable subject matter." This decision spawned a wave of new biotechnology firms and the infant industry's first investment boom. In 1982 recombinant insulin became the first product made through genetic engineering to secure approval from the U.S. Food and Drug Administration (FDA). Since then, dozens of genetically engineered protein medications have been commercialized around the world, including recombinant versions of growth hormone, clotting factors, proteins for stimulating the production of red and white blood cells, interferons, and clot-dissolving agents.

In the early years, the main achievement of biotechnology was the ability to produce naturally occurring therapeutic molecules in larger quantities than could be derived from conventional sources such as plasma, animal organs, and human cadavers. Recombinant proteins are also less likely to be contaminated with pathogens or to provoke allergic reactions. Today, biotechnology researchers seek to discover the root molecular causes of disease and to intervene precisely at that level. Sometimes this means producing therapeutic proteins that augment the body's own supplies or that make up for genetic deficiencies, as in the first generation of biotech medications. (Gene therapy—insertion of genes encoding a needed protein into a patient's body or cells—is a related approach.) But the biotechnology industry has also expanded its research into the development of traditional pharmaceuticals and monoclonal antibodies that stop the progress of a disease. Such steps are uncovered through painstaking study of genes (genomics), the proteins that they encode (proteomics), and the larger biological pathways in which they act.

In addition to the tools mentioned above, biotechnology also involves merging biological information with computer technology (bioinformatics), exploring the use of microscopic equipment that can enter the human body (nanotechnology), and possibly applying techniques of stem cell research and cloning to replace dead or defective cells and tissues (regenerative medicine). Companies and academic laboratories integrate these disparate technologies in an effort to analyze downward into molecules and also to synthesize upward from molecular biology toward chemical pathways, tissues, and organs.

In addition to being used in health care, biotechnology has proved helpful in refining industrial processes through the discovery and production of biological enzymes that spark chemical reactions (catalysts); for environmental cleanup, with enzymes that digest contaminants into harmless chemicals and then die after consuming the available "food supply"; and in agricultural production through genetic engineering.

Agricultural applications of biotechnology have proved the most controversial. Some activists and consumer groups have called for bans on genetically modified organisms (GMOs) or for labeling laws to inform consumers of the growing presence of GMOs in the food supply. In the United States, the introduction of GMOs into agriculture began in 1993, when the FDA approved bovine somatotropin (BST), a growth hormone that boosts milk production in dairy cows. The next year, the FDA approved the first genetically modified whole food, a tomato engineered for a longer shelf life. Since then, regulatory approval in the United States, Europe, and elsewhere has been won by dozens of agricultural GMOs, including crops that produce their own pesticides and crops that survive the application of specific herbicides used to kill weeds. Studies by the United Nations, the U.S. National Academy of Sciences, the European Union, the American Medical Association, U.S. regulatory agencies, and other organizations have found GMO foods to be safe, but skeptics contend that it is still too early to judge the long-term health and ecological effects of such crops. In the late 20th and early 21st centuries, the land area planted in genetically modified crops increased dramatically, from 1.7 million hectares (4.2 million acres) in 1996 to 160 million hectares (395 million acres) by 2011.

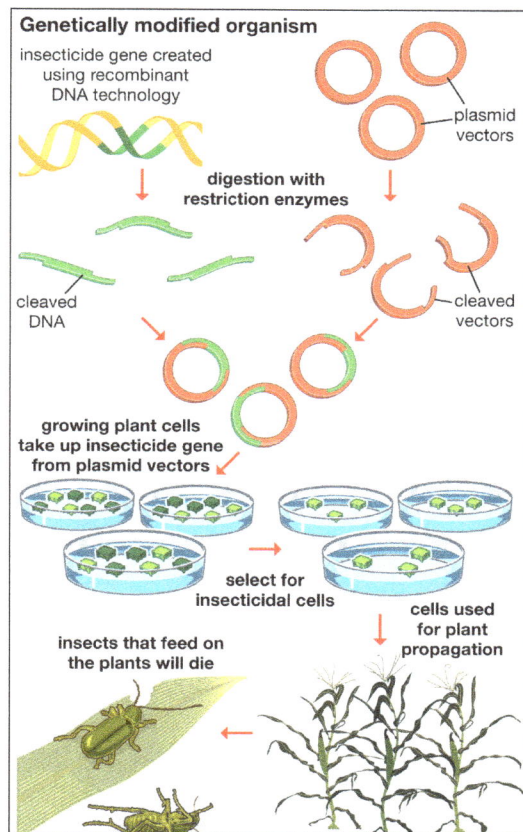

Genetically modified organisms are produced using scientific methods that include recombinant DNA technology.

Overall, the revenues of U.S. and European biotechnology industries roughly doubled over the five-year period from 1996 through 2000. Rapid growth continued into the 21st century, fueled by the introduction of new products, particularly in health care.

Contrary to its name, biotechnology is not a single technology. Rather it is a group of technologies that share two (common) characteristics -- working with living cells and their molecules and having a wide range of practice uses that can improve our lives.

Biotechnology can be broadly defined as "using organisms or their products for commercial purposes." As such, (traditional) biotechnology has been practices since he beginning of records history. (It has been used to:) bake bread, brew alcoholic beverages, and breed food crops or domestic animals. But recent developments in molecular biology have given biotechnology new meaning, new prominence, and new potential. It is (modern) biotechnology that has captured the attention of the public. Modern biotechnology can have a dramatic effect on the world economy and society

One example of modern biotechnology is genetic engineering. Genetic engineering is the process of transferring individual genes between organisms or modifying the genes in an organism to remove or add a desired trait or characteristic. Examples of genetic engineering are described later in this document. Through genetic engineering, genetically modified crops or organisms are formed. These GM crops or GMOs are used to produce biotech-derived foods. It is this specific type of modern biotechnology, genetic engineering, that seems to generate the most attention and concern by consumers and consumer groups. What is interesting is that modern biotechnology is far more precise than traditional forms of biotechnology and so is viewed by some as being far safer.)

Working of Modern Biotechnology

All organisms are made up of cells that are programmed by the same basic genetic material, called DNA (deoxyribonucleic acid). Each unit of DNA is made up of a combination of the following nucleotides -- adenine (A), guanine (G), thymine (T), and cytosine (D) -- as well as a sugar and a phosphate. These nucleotides pair up into strands that twist together into a spiral structure call a "double helix." This double helix is DNA. Segments of the DNA tell individual cells how to produce specific proteins. These segments are genes. It is the presence or absence of the specific protein that gives an organism a trait or characteristic. More than 10,000 different genes are found in most plant and animal species. This total set of genes for an organism is organized into chromosomes within the cell nucleus. The process by which a multicellular organism develops from a single cell through an embryo stage into an adult is ultimately controlled by the genetic information of the cell, as well as interaction of genes and gene products with environmental factors.

When cells reproduce, the DNA strands of the double helix separate. Because nucleotide A always pairs with T and G always pairs with C, each DNA strand serves as a precise blueprint for a specific protein. Except for mutations or mistakes in the replication process, a single cell is equipped with the information to replicate into millions of identical cells. Because all organisms are made up of the same type of genetic material (nucleotides A, T, G, and C), biotechnologists use enzymes to cut and remove DNA segments from one organism and recombine it with DNA in another organism. This is called recombinant DNA (rDNA) technology, and it is one of the basic tools of modern biotechnology. rDNA technology is the laboratory manipulation of DNA in which DNA, or fragments of DNA from different sources, are cut and recombined using enzymes. This recombinant DNA is then inserted into a living organism. rDNA technology is usually used synonymously with genetic engineering. rDNA technology allows researchers to move genetic information between unrelated organisms to produce desired products or characteristics or to eliminate undesirable characteristics.

Genetic engineering is the technique of removing, modifying or adding genes to a DNA molecule in order to change the information it contains. By changing this information, genetic engineering changes the type or amount of proteins an organism is capable of producing. Genetic engineering is used in the production of drugs, human gene therapy, and the development of improved plants. For example, an "insect protection" gene (Bt) has been inserted into several crops - corn, cotton, and potatoes - to give farmers new tools for integrated pest management. Bt corn is resistant to European corn borer. This inherent resistance thus reduces a farmers pesticide use for controlling European corn borer, and in turn requires less chemicals and potentially provides higher yielding Agricultural Biotechnology.

Although major genetic improvements have been made in crops, progress in conventional breeding programs has been slow. In fact, most crops grown in the US produce less than their full genetic potential. These shortfalls in yield are due to the inability of crops to tolerate or adapt to environmental stresses, pests, and diseases. For example, some of the world's highest yields of potatoes are in Idaho under irrigation, but in 1993 both quality and yield were severely reduced because of cold, wet weather and widespread frost damage during June. Some of the world's best bread wheats and malting barleys are produced in the north-central states, but in 1993 the disease Fusarium caused an estimated $1 billion in damage.

Scientists have the ability to insert genes that give biological defense against diseases and insects, thus reducing the need for chemical pesticides, and they will soon be able to convey genetic traits that enable crops to better withstand harsh conditions, such as drought.

In 1995, ILTAB reported the first transfer through biotechnology of a resistance gene from a wild species of rice to a susceptible cultivated rice variety. The transferred gene expressed resistance to Xanthomonas oryzae, a bacterium which can destroy the crop through disease. The resistant gene was transferred into susceptible rice varieties that are cultivated on more than 24 million hectares around the world.

Benefits can also be seen in the environment, where insect-protected biotech crops reduce the need for chemical pesticide use. Insect-protected crops allow for less potential exposure of farmers and groundwater to chemical residues, while providing farmers with season-long control. Also by reducing the need for pest control, impacts and resources spent on the land are less, thereby preserving the topsoil.

Major advances also have been made through conventional breeding and selection of livestock, but significant gains can still be made by using biotechnology. Currently, farmers in the U.S spend $17 billion dollars on animal health. Diseases such as hog cholera and pests such as screwworm have been eradicated. Uses of biotechnology in animal production include development of vaccines to protect animals from disease, production of several calves from one embryo (cloning), increase of animal growth rate, and rapid disease detection.

Modern biotechnology has offered opportunities to produce more nutritious and better tasting foods, higher crop yields and plants that are naturally protected from disease and insects. Modern biotechnology allows for the transfer of only one or a few desirable genes, thereby permitting scientists to develop crops with specific beneficial traits and reduce undesirable traits.

Traditional biotechnology such as cross-pollination in corn produces numerous, non-selective changes. Genetic modifications have produced fruits that can ripen on the vine for better taste, yet have longer shelf lives through delayed pectin degradation. Tomatoes and other produce containing increased levels of certain nutrients, such as vitamin C, vitamin E, and or beta carotene, and help protect against the risk of chronic diseases, such as some cancers and heart disease. Similarly introducing genes that increase available iron levels in rice three-fold is a potential remedy for iron deficiency, a condition that effects more than two billion people and causes anemia in about half that number. Most of the today's hard cheese products are made with a biotech enzyme called chymosin. This is produced by genetically engineered bacteria which is considered more purer and plentiful than it's naturally occurring counterpart, rennet, which is derived from calf stomach tissue.

In 1992, Monsanto Company successfully inserted a gene from a bacterium into the Russet Burbank potato. This gene increases the starch content of the potato. Higher starch content reduces oil absorption during frying, thereby lowering the cost of processing french fries and chips and reducing the fat content in the finished product. This product is still awaiting final development and approval.

Modern biotechnology offers effective techniques to address food safety concerns. Biotechnical methods may be used to decrease the time necessary to detect foodborne pathogens, toxins, and chemical contaminants, as well as to increase detection sensitivity. Enzymes, antibodies, and microorganisms produced using rDNA techniques are being used to monitor food production and processing systems for quality control.

Biotechnology can compress the time frame required to translate fundamental discoveries into applications. This is done by controlling which genes are altered in an organized fashion. For example, a known gene sequence from a corn plant can be altered to improve yield, increase drought tolerance, and produce insect resistance (Bt) in one generation. Conventional breeding techniques would take several years. Conventional breeding techniques would require that a field of corn is grown and each trait is selected from individual stalks of corn. The ears of corn from selected stalks with each desired trait (e.g, drought tolerance and yield performance) would then be grown and combined (cross-pollinated). Their offspring (hybrid) would be further selected for the desired result (a high performing corn with drought tolerance). With improved technology and knowledge about agricultural organisms, processes, and ecosystems, opportunities will emerge to produce new and improved agricultural products in an environmentally sound manner.

In summary, modern biotechnology offers opportunities to improve product quality, nutritional content, and economic benefits. The genetic makeup of plants and animals can be modified by either insertion of new useful genes or removal of unwanted ones. Biotechnology is changing the way plants and animals are grown, boosting their value to growers, processors, and consumers.

Industrial Biotechnology

Industrial biotechnology applies the techniques of modern molecular biology to improve the efficiency and reduce the environmental impacts of industrial processes like textile, paper and pulp, and chemical manufacturing. For example, industrial biotechnology companies develop biocatalysts, such as enzymes, to synthesize chemicals. Enzymes are proteins produced by all organisms. Using biotechnology, the desired enzyme can be manufactured in commercial quantities.

Commodity chemicals (e.g., polymer-grade acrylamide) and specialty chemicals can be produced using biotech applications. Traditional chemical synthesis involves large amounts of energy and often-undesirable products, such as HCl. Using biocatalysts, the same chemicals can be produced more economically and more environmentally friendly. An example would be the substitution of protease in detergents for other cleaning compounds. Detergent proteases, which remove protein impurities, are essential components of modern detergents. They are used to break down protein, starch, and fatty acids present on items being washed. Protease production results in a biomass that in turn yields a useful byproduct- an organic fertilizer. Biotechnology is also used in the textile industry for the finishing of fabrics and garments. Biotechnology also produces biotech-derived cotton that is warmer, stronger, has improved dye uptake and retention, enhanced absorbency, and wrinkle- and shrink-resistance.

Some agricultural crops, such as corn, can be used in place of petroleum to produce chemicals. The crop's sugar can be fermented to acid, which can be then used as an intermediate to produce other chemical feedstocks for various products. It has been projected that 30% of the world's chemical and fuel needs could be supplied by such renewable resources in the first half of the next century. It has been demonstrated, at test scale, that biopulping reduces the electrical energy required for wood pulping process by 30%.

Environmental Biotechnology

Environmental biotechnology is the used in waste treatment and pollution prevention. Environmental biotechnology can more efficiently clean up many wastes than conventional methods and greatly reduce our dependence on methods for land-based disposal.

Every organism ingests nutrients to live and produces by-products as a result. Different organisms need different types of nutrients. Some bacteria thrive on the chemical components of waste products. Environmental engineers use bioremediation, the broadest application of environmental biotechnology, in two basic ways. They introduce nutrients to stimulate the activity of bacteria already present in the soil at a waste site, or add new bacteria to the soil. The bacteria digest the waste at the site and turn it into harmless byproducts. After the bacteria consume the waste materials, they die off or return to their normal population levels in the environment.

Bioremediation, is an area of increasing interest. Through application of biotechnical methods, enzyme bioreactors are being developed that will pretreat some industrial waste and food waste components and allow their removal through the sewage system rather than through solid waste disposal mechanisms. Waste can also be converted to biofuel to run generators. Microbes can be induced to produce enzymes needed to convert plant and vegetable materials into building blocks for biodegradable plastics.

In some cases, the byproducts of the pollution-fighting microorganisms are themselves useful. For example, methane can be derived from a form of bacteria that degrades sulfur liquor, a waste product of paper manufacturing. This methane can then be used as a fuel or in other industrial processes.

Human Applications

Biotechnical methods are now used to produce many proteins for pharmaceutical and other specialized purposes. A harmless strain of Escherichia coli bacteria, given a copy of the gene for human insulin, can make insulin. As these genetically modified (GM) bacterial cells age, they produce human insulin, which can be purified and used to treat diabetes in humans.

Microorganisms can also be modified to produce digestive enzymes. In the future, these microorganisms could be colonized in the intestinal tract of persons with digestive enzyme insufficiencies. Products of modern biotechnology include artificial blood vessels from collagen tubes coated with a layer of the anticoagulant heparin.

Gene therapy – altering DNA within cells in an organism to treat or cure a disease – is one of the most promising areas of biotechnology research. New genetic therapies are being developed to treat diseases such as cystic fibrosis, AIDS and cancer.

DNA fingerprinting is the process of cross matching two strands of DNA. In criminal investigations, DNA from samples of hair, bodily fluids or skin at a crime scene are compared with those obtained from the suspects. In practice, it has become one of the most powerful and widely known applications of biotechnology today. Another process, polymerase chain reaction (PCR), is also being used to more quickly and accurately identify the presence of infections such as AIDS, Lyme disease and Chlamydia.

Paternity determination is possible because a child's DNA pattern is inherited, half from the mother and half from the father. To establish paternity, DNA fingerprints of the mother, child and the alleged father are compared. The matching sequences of the mother and the child are eliminated from the child's DNA fingerprint; what remains comes from the biological father.

These segments are then compared for a match with the DNA fingerprint of the alleged father. DNA testing is also used on human fossils to determine how closely related fossil samples are from different geographic locations and geologic areas. The results shed light on the history of human evolution and the manner in which human ancestors settled different parts of the world.

Biotechnology for the 21st Century

Experts in United States anticipate the world's population in 2050 to be approximately 8.7 billion persons. The world's population is growing, but its surface area is not. Compounding the effects of population growth is the fact that most of the earth's ideal farming land is already being utilized. To avoid damaging environmentally sensitive areas, such as rain forests, we need to increase crop yields for land currently in use. By increasing crop yields, through the use of biotechnology the constant need to clear more land for growing food is reduced.

Countries in Asia, Africa, and elsewhere are grappling with how to continue feeding a growing population. They are also trying to benefit more from their existing resources. Biotechnology holds the key to increasing the yield of staple crops by allowing farmers to reap bigger harvests from currently cultivated land, while preserving the land's ability to support continued farming.

Malnutrition in underdeveloped countries is also being combated with biotechnology. The Rockefeller Foundation is sponsoring research on "golden rice", a crop designed to improve nutrition in the developing world. Rice breeders are using biotechnology to build Vitamin A into the rice. Vitamin A deficiency is a common problem in poor countries. A second phase of the project will increase the iron content in rice to combat anemia, which is widespread problem among women and children in underdeveloped countries. Golden rice, expected to be for sale in Asia in less than five years, will offer dramatic improvements in nutrition and health for millions of people, with little additional costs to consumers.

Similar initiatives using genetic manipulation are aimed at making crops more productive by reducing their dependence on pesticides, fertilizers and irrigation, or by increasing their resistance to plant diseases.

Increased crop yield, greater flexibility in growing environments, less use of chemical pesticides and improved nutritional content make agricultural biotechnology, quite literally, the future of the world's food supply.

Concerns about Biotechnology

As biotechnology has become widely used, questions and concerns have also been raised. The most vocal opposition has come from European countries. One of the main areas of concern is the safety of genetically engineered food.

In assessing the benefits and risks involved in the use of modern biotechnology, there are a series of issues to be addressed so that informed decisions can be made. In making value judgments about risks and benefits in the use of biotechnology, it is important to distinguish between technology-inherent risks and technology-transcending risks. The former includes assessing any risks associated with food safety and the behavior of a biotechnology-based product in the environment. The latter involve the political and social context in which the technology is used, including how these uses may benefit or harm the interests of different groups in society.

The health effects of foods grown from genetically engineered crop depend on the composition of the food itself. Any new product may have either beneficial or occasional harmful effects on human health. For example, a biotech-derived food with a higher content of digestible iron is likely to have a positive effect if consumed by iron-deficient individuals. Alternatively, the transfer of genes from one species to another may also transfer the risk for exposure to allergens. These risks are systematically evaluated by FDA and identified prior to commercialization.

Individuals allergic to certain nuts, for example, need to know if genes conveying this trait are transferred to other foods such as soybeans. Labeling would be required if such crops were available to consumers.

Among the potential ecological risks identified are increased weediness, due to cross- pollination from genetically modified crops spreads to other plants in nearby fields. This may allow the spread of traits such as herbicide-resistance to non-target plants that could potentially develop into weeds. This ecological risk is assessed when deciding if a plant with a given trait should be released into a particular environment, and if so, under what conditions.

Other potential ecological risks stem from the use of genetically modified corn and cotton with insecticidal genes from Bacillus thuringiensis (Bt genes). This may lead to the development of resistance to Bt in insect populations exposed to the biotech-derived crop. There also may be risks to non-target species, such as birds and butterflies, from the plants with Bt genes. The monitoring of these effects of new crops in the environment and implementation of effective risk management approaches is an essential component of further research. It is also important to keep all risks in perspective by comparing the products of biotechnology and conventional agriculture.

The reduction of biodiversity would represent a technology-transcending risk. Reduced biological diversity due to destruction of tropical forests, conversion of land to agriculture, overfishing, and the other practices to feed a growing world population is a significant loss far more than any potential loss of biodiversity due to biotech-derived crop varieties. Improved governance and international support are necessary to limit loss of biodiversity.

What we know from our understanding of science and more than a decade of experience with biotech-derived plants is the following: There is no evidence that genetic transfers between unrelated organisms pose human health concerns that are different from those encountered with any new plant or animal variety. The risks associated with biotechnology are the same as those associated with plants and microbes developed by conventional methods.

Branches of Biotechnology

Biotechnology has more than a few different branches which are referred to by dissimilar terms mainly noticeable with dissimilar colors to clarify the biotechnological field that it is used in. First of all there is the red biotechnology that is used for medical processes, like finding genetic cures by going through genomic manipulations and creating organisms to produce antibiotics.

Green biotechnology is used in orientation to agricultural process that use biotechnology. Some examples of that would be the development of transgenic plants that are calculated to survive under precise environmental circumstances. A large goal of the green biotechnology is to expand more environment friendly solutions, for example to find a way take out the need for pesticides.

Green biotechnology which is more usually known as Plant Biotechnology is a rapidly increasing field within Modern biotechnology. It essentially involves the opening of foreign genes into inexpensively important plant species, resulting in crop improvement and the production of novel products in plants.

The term used for manufacturing biotechnology is white biotechnology. This type of biotechnology is used to decrease the costs for producing industrial supplies that occur when traditional processes are used. For example, white biotechnology can expand an organism that is capable to produce a certain useful chemical by natural processes quite than by industrial ways that it was done earlier.

The industrial biotechnology society usually accepts an informal divide between manufacturing and pharmaceutical biotechnology. An example would be that of company rising fungus to produce antibiotics, e.g. penicillin from the penicillium fungi.

Some additional examples of branches of biotechnology are blue biotechnologies that deal with marine and a marine usage of biotechnology, but that is not very extensively used. When discussing about not the straight research part of biotechnology then bioeconomy is used to talk about the savings and the economical benefits that biotechnology brings.

Biotechnology with marine organisms, feed into aquaculture, marine animal and fish health, marine natural goods (including medicines), biofilms, bioremediation, marine ecology and bio-oceanography and other marine products (e.g. enzymes).

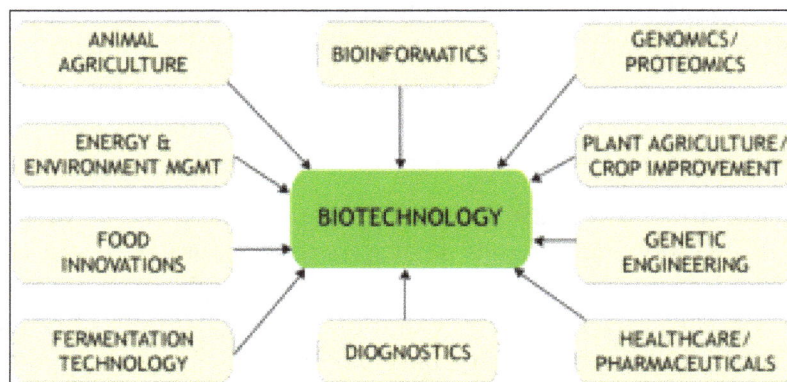

There are several branches of Biotechnology. They are classified as:

Blue Biotechnology

This branch of biotechnology helps to control the marine organisms and water borne organisms. It is a process which has to do with marine or underwater environment. The use of this biotechnology is very rare. Blue biotechnology is used to protect the marine organisms from harmful diseases underwater.

Bioinformatics

Bioinformatics is the combination of computer and biotechnology. It helps in finding the analysis of datas related to Biotechnology. It is used for various purposes like drugs, for the development of medicines; it is also used to improve the fertility of crops and plants and also for pest, drought and it is resistance to diseases. Bioinformatics is known and referred by the term computational biology. It plays an important and a vital role in areas like Functional genomics, structural genomics and proteomics these areas contribute a lot and become a key contributor to Biotechnology and pharma sector.

Green Biotechnology

Green Biotechnology is the term used for the agricultural sector. With the help of the process called the Micropropagation (a practice of producing larger number of plants through the existing stock of plants) which helps in selecting the right quality of plants and crops. Also with the help of Transgenic plants (plants whose DNA is modified); this design of transgenic plants helps to grow in a specified environment with the help of certain chemicals.

Red Biotechnology

Red biotechnology is referred to as Medical Biotechnology. It is used for the production of drugs and antibiotic medicines. It also helps to create or design organisms. Through the process of genetic manipulation it helps to cure genetic issues in organisms. It also helps in analysing diseases in organisms. It also helps in developing new ways of diagnosis by performing tests. With the help of stem cell therapy it helps the organs to grow and it also cures the damaged issues in organisms.

White Biotechnology

White Biotechnology is also called and known by the name Industry Biotechnology. This kind of biotechnology is used and applied in industries and its processes. The various uses of this Biotechnology includes; biopolymers (Plastics) Substitutes, new invention of vehicle parts and fuels for the vehicles, invention of fibres for the clothing industry, it is also involved in developing new chemicals and the production process.

Biological Engineering

Biological engineering, or bioengineering/bio-engineering, is the application of principles of biology and the tools of engineering to create usable, tangible, economically viable products. Biological engineering employs knowledge and expertise from a number of pure and applied sciences, such as mass and heat transfer, kinetics, biocatalysts, biomechanics, bioinformatics, separation and purification processes, bioreactor design, surface science, fluid mechanics, thermodynamics, and polymer science. It is used in the design of medical devices, diagnostic equipment, biocompatible materials, renewable bioenergy, ecological engineering, agricultural engineering, and other areas that improve the living standards of societies. Examples of bioengineering research include bacteria engineered to produce chemicals, new medical imaging technology, portable and rapid disease diagnostic devices, prosthetics, biopharmaceuticals, and tissue-engineered organs. Bioengineering overlaps substantially with biotechnology and the biomedical sciences in a way analogous to how various other forms of engineering and technology relate to various other sciences (for example, aerospace engineering and other space technology to kinetics and astrophysics).

In general, biological engineers (or *biomedical engineers*) attempt to either mimic biological systems to create products or modify and control biological systems so that they can replace, augment, sustain, or predict chemical and mechanical processes. Bioengineers can apply their expertise to other applications of engineering and biotechnology, including genetic modification of plants and microorganisms, bioprocess engineering, and biocatalysis. Working with doctors, clinicians and researchers, bioengineers use traditional engineering principles and techniques and apply them to real-world biological and medical problems.

Biological engineering is a science-based discipline founded upon the biological sciences in the same way that chemical engineering, electrical engineering, and mechanical engineering can be based upon chemistry, electricity and magnetism, and classical mechanics, respectively.

Before WWII, biological engineering had just begun being recognized as a branch of engineering, and was a very new concept to people. Post-WWII, it started to grow more rapidly, partially due to the term "bioengineering" being coined by British scientist and broadcaster Heinz Wolff in 1954 at the National Institute for Medical Research. Wolff graduated that same year and became the director of the Division of Biological Engineering at the university. This was the first time Bioengineering was recognized as its own branch at a university. Electrical engineering is considered to pioneer this engineering sector due to its work with medical devices and machinery during this time.When engineers and life scientists started working together, they recognized the problem that the engineers didn't know enough about the actual biology behind their work. To resolve this problem, engineers who wanted to get into biological engineering devoted more of their time and studies to the details and processes that go into fields such as biology, psychology, and medicine. The term biological engineering may also be applied to environmental modifications such as surface soil protection, slope stabilization, watercourse and shoreline protection, windbreaks, vegetation barriers including noise barriers and visual screens, and the ecological enhancement of an area. Because other engineering disciplines also address living organisms, the term biological engineering can be applied more broadly to include agricultural engineering.

The first biological engineering program was created at University of California, San Diego in 1966, making it the first biological engineering curriculum in the United States. More recent programs have been launched at MIT and Utah State University. Many old agricultural engineering departments in universities over the world have re-branded themselves as *agricultural and biological engineering* or *agricultural and biosystems engineering,* due to biological engineering as a whole being a rapidly developing field with fluid categorization. According to Professor Doug

Lauffenburger of MIT, biological engineering has a broad base which applies engineering principles to an enormous range of size and complexities of systems. These systems range from the molecular level (molecular biology, biochemistry, microbiology, pharmacology, protein chemistry, cytology, immunology, neurobiology and neuroscience) to cellular and tissue-based systems (including devices and sensors), to whole macroscopic organisms (plants, animals), and can even range up to entire ecosystems.

Sub-disciplines

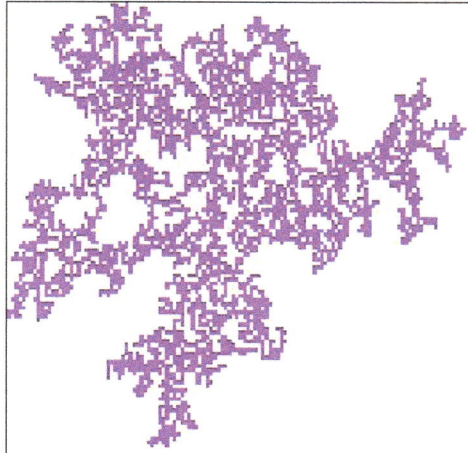

Modeling of the spread of disease using Cellular Automata and Nearest Neighbor Interactions.

Depending on the institution and particular definitional boundaries employed, some major branches of bioengineering may be categorized as (note these may overlap):

- Biomedical engineering: Application of engineering principles and design concepts to medicine and biology for healthcare purposes:

  ◦ Tissue engineering

  ◦ Genetic engineering

  ◦ Neural engineering

  ◦ Pharmaceutical engineering

  ◦ Clinical engineering

  ◦ Bioinformatics

  ◦ Biomechanics

- Biochemical engineering: Fermentation engineering, application of engineering principles to microscopic biological systems that are used to create new products by synthesis, including the production of protein from suitable raw materials.

- Biological systems engineering: Application of engineering principles and design concepts to agriculture, food sciences, and ecosystems.

- Bioprocess engineering: Develops technology to monitor the conditions of the where the process of making pharmaceuticals takes place, (Ex: bioprocess design, biocatalysis, bioseparation, bioinformatics, bioenergy).

- Environmental health engineering: Application of engineering principles to the control of the environment for the health, comfort, and safety of human beings. It includes the field of life-support systems for the exploration of outer space and the ocean.

- Human-factors engineering: Application of engineering, physiology, and psychology to the optimization of the human–machine relationship.

- Biotechnology: The use of living systems and organisms to develop or make products. (Ex: pharmaceuticals).

- Biomimetics: The imitation of models, systems, and elements of nature for the purpose of solving complex human problems. (Ex: velcro, designed after George de Mestral noticed how easily burs stuck to a dog's hair).

- Bioelectrical engineering.

- Biomechanical engineering.

- Bionics: An integration of Biomedical, focused more on the robotics and assisted technologies. (Ex: prosthetics).

- Bioprinting: Utilizing biomaterials to print organs and new tissues.

- Biorobotics: (Ex: prosthetics).

- Systems biology: The study of biological systems.

References

- Biotechnology, technology: britannica.com, Retrieved 12 July, 2019

- Abramovitz, melissa (2015). Biological engineering. Gale virtual reference library. P. 10. Isbn 978-1-62968-526-7

- Biotech-applications, documents, extension-program: ncsu.edu, Retrieved 11 January, 2019

- herold, keith; bentley, william e.; vossoughi, jafar (2010). The basics of bioengineering education. 26th southern biomedical engineering conference, college park, maryland. P. 65. Isbn 9783642149979

- Branches-of-biotech: biotechonweb.com, Retrieved 1 August, 2019

- "biotechnology vs biomedical science vs biomedical engineering (bioengineering)". Tanmoy ray. 2018-07-19. Retrieved 2018-07-21

- Abramovitz, melissa (2015). Biological engineering. Gale virtual reference library. P. 18. Isbn 978-1-62968-526-7

2
Concepts in Biotechnology

The fundamental concepts of biotechnology are related to cell, gene, proteins, DNA, RNA, fermentation and the human genome project. The topics will help in gaining a better perspective about these concepts in biotechnology.

Cell

Cell is the basic membrane-bound unit that contains the fundamental molecules of life and of which all living things are composed. A single cell is often a complete organism in itself, such as a bacterium or yeast. Other cells acquire specialized functions as they mature. These cells cooperate with other specialized cells and become the building blocks of large multicellular organisms, such as humans and other animals. Although cells are much larger than atoms, they are still very small. The smallest known cells are a group of tiny bacteria called mycoplasmas; some of these single-celled organisms are spheres as small as 0.2 μm in diameter (1μm = about 0.000039 inch), with a total mass of 10–14 gram—equal to that of 8,000,000,000 hydrogen atoms. Cells of humans typically have a mass 400,000 times larger than the mass of a single mycoplasma bacterium, but even human cells are only about 20 μm across. It would require a sheet of about 10,000 human cells to cover the head of a pin, and each human organism is composed of more than 75,000,000,000,000 cells.

As an individual unit, the cell is capable of metabolizing its own nutrients, synthesizing many types of molecules, providing its own energy, and replicating itself in order to produce succeeding generations. It can be viewed as an enclosed vessel, within which innumerable chemical reactions take place simultaneously. These reactions are under very precise control so that they contribute to the life and procreation of the cell. In a multicellular organism, cells become specialized to perform different functions through the process of differentiation. In order to do this, each cell keeps in constant communication with its neighbours. As it receives nutrients from and expels wastes into its surroundings, it adheres to and cooperates with other cells. Cooperative assemblies of similar cells form tissues, and a cooperation between tissues in turn forms organs, which carry out the functions necessary to sustain the life of an organism.

The Nature and Function of Cells

A cell is enclosed by a plasma membrane, which forms a selective barrier that allows nutrients to enter and waste products to leave. The interior of the cell is organized into many specialized compartments, or organelles, each surrounded by a separate membrane. One major organelle, the nucleus, contains the genetic information necessary for cell growth and reproduction. Each cell contains only one nucleus, whereas other types of organelles are present in multiple copies in the cellular contents, or cytoplasm. Organelles include mitochondria, which are responsible for the energy transactions necessary for cell survival; lysosomes, which digest unwanted materials within the cell; and the endoplasmic reticulum and the Golgi apparatus, which play important roles in the internal organization of the cell by synthesizing selected molecules and then processing, sorting, and directing them to their proper locations. In addition, plant cells contain chloroplasts, which are responsible for photosynthesis, whereby the energy of sunlight is used to convert molecules of carbon dioxide (CO_2) and water (H_2O) into carbohydrates. Between all these organelles is the space in the cytoplasm called the cytosol. The cytosol contains an organized framework of fibrous molecules that constitute the cytoskeleton, which gives a cell its shape, enables organelles to move within the cell, and provides a mechanism by which the cell itself can move. The cytosol also contains more than 10,000 different kinds of molecules that are involved in cellular biosynthesis, the process of making large biological molecules from small ones.

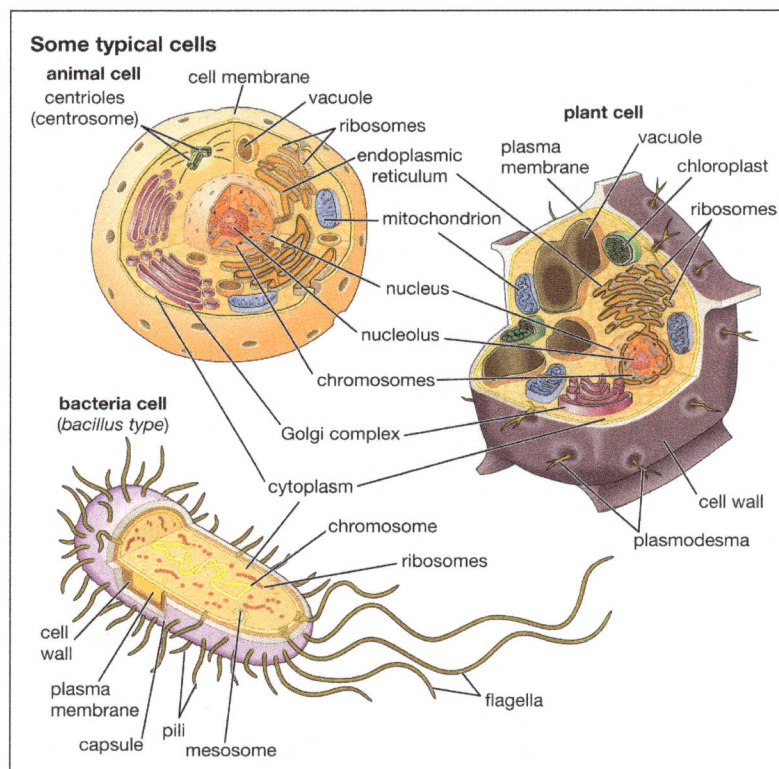

Animal cells and plant cells contain membrane-bound organelles, including a distinct nucleus.
In contrast, bacterial cells do not contain organelles.

Specialized organelles are a characteristic of cells of organisms known as eukaryotes. In contrast, cells of organisms known as prokaryotes do not contain organelles and are generally smaller than eukaryotic cells. However, all cells share strong similarities in biochemical function.

Animal cell

lysosome   ribosomes
centriole   cilium
centrosome   cell membrane
peroxisome   smooth
                                   endoplasmic
                                   reticulum
                            nuclear pore
                            nucleolus
                            nucleoplasm
                            nuclear
                            envelope
                                  rough
                                  endoplasmic
                                  reticulum
Golgi
apparatus   mitochondrion
secretory vesicles   cytoplasm

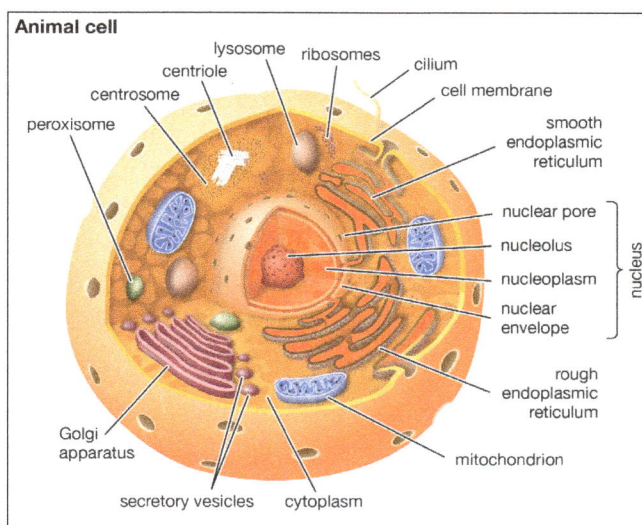

Cutaway drawing of a eukaryotic cell.

The Molecules of Cells

Cells contain a special collection of molecules that are enclosed by a membrane. These molecules give cells the ability to grow and reproduce. The overall process of cellular reproduction occurs in two steps: cell growth and cell division. During cell growth, the cell ingests certain molecules from its surroundings by selectively carrying them through its cell membrane. Once inside the cell, these molecules are subjected to the action of highly specialized, large, elaborately folded molecules called enzymes. Enzymes act as catalysts by binding to ingested molecules and regulating the rate at which they are chemically altered. These chemical alterations make the molecules more useful to the cell. Unlike the ingested molecules, catalysts are not chemically altered themselves during the reaction, allowing one catalyst to regulate a specific chemical reaction in many molecules.

Biological catalysts create chains of reactions. In other words, a molecule chemically transformed by one catalyst serves as the starting material, or substrate, of a second catalyst and so on. In this way, catalysts use the small molecules brought into the cell from the outside environment to create increasingly complex reaction products. These products are used for cell growth and the replication of genetic material. Once the genetic material has been copied and there are sufficient molecules to support cell division, the cell divides to create two daughter cells. Through many such cycles of cell growth and division, each parent cell can give rise to millions of daughter cells, in the process converting large amounts of inanimate matter into biologically active molecules.

The Structure of Biological Molecules

Cells are largely composed of compounds that contain carbon. The study of how carbon atoms interact with other atoms in molecular compounds forms the basis of the field of organic chemistry and plays a large role in understanding the basic functions of cells. Because carbon atoms can form stable bonds with four other atoms, they are uniquely suited for the construction of complex molecules. These complex molecules are typically made up of chains and rings that contain hydrogen, oxygen, and nitrogen atoms, as well as carbon atoms. These molecules may consist of anywhere from 10 to millions of atoms linked together in specific arrays. Most, but not all, of the carbon-containing molecules in cells are built up from members of one of four different families of small organic molecules: sugars, amino acids, nucleotides, and fatty acids. Each of these families contains a group of molecules that resemble one another in both structure and function. In addition to other important functions, these molecules are used to build large macromolecules. For example, the sugars can be linked to form polysaccharides such as starch and glycogen, the amino acids can be linked to form proteins, the nucleotides can be linked to form the DNA (deoxyribonucleic acid) and RNA (ribonucleic acid) of chromosomes, and the fatty acids can be linked to form the lipids of all cell membranes.

Aside from water, which forms 70 percent of a cell's mass, a cell is composed mostly of macromolecules. By far the largest portion of macromolecules are the proteins. An average-sized protein macromolecule contains a string of about 400 amino acid molecules. Each amino acid has a different side chain of atoms that interact with the atoms of side chains of other amino acids. These interactions are very specific and cause the entire protein molecule to fold into a compact globular form. In theory, nearly an infinite variety of proteins can be formed, each with a different sequence of amino acids. However, nearly all these proteins would fail to fold in the unique ways required to form efficient functional surfaces and would therefore be useless to the cell. The proteins present in cells of modern animals and humans are products of a long evolutionary history, during which the ancestor proteins were naturally selected for their ability to fold into specific three-dimensional forms with unique functional surfaces useful for cell survival.

Approximate chemical composition of a typical mammalian cell	
component	percent of total cell weight
water	70
inorganic ions (sodium, potassium, magnesium, calcium, chloride, etc.)	1
miscellaneous small metabolites	3
proteins	18
RNA	1.1
DNA	0.25
phospholipids and other lipids	5
polysaccharides	2

Most of the catalytic macromolecules in cells are enzymes. The majority of enzymes are proteins. Key to the catalytic property of an enzyme is its tendency to undergo a change in its shape when it binds to its substrate, thus bringing together reactive groups on substrate molecules. Some enzymes are macromolecules of RNA, called ribozymes. Ribozymes consist of linear chains of nucleotides that fold in specific ways to form unique surfaces, similar to the ways in which proteins fold. As with proteins, the specific sequence of nucleotide subunits in an RNA chain gives each macromolecule a unique character. RNA molecules are much less frequently used as catalysts in cells than are protein molecules, presumably because proteins, with the greater variety of amino acid side chains, are more diverse and capable of complex shape changes. However, RNA molecules are thought to have preceded protein molecules during evolution and to have catalyzed most of the chemical reactions required before cells could evolve.

Coupled Chemical Reactions

Cells must obey the laws of chemistry and thermodynamics. When two molecules react with each other inside a cell, their atoms are rearranged, forming different molecules as reaction products and releasing or consuming energy in the process. Overall, chemical reactions occur only in one direction; that is, the final reaction product molecules cannot spontaneously react, in a reversal of the original process, to reform the original molecules. This directionality of chemical reactions is explained by the fact that molecules only change from states of higher free energy to states of lower free energy. Free energy is the ability to perform work (in this case, the "work" is the rearrangement of atoms in the chemical reaction). When work is performed, some free energy is used and lost, with the result that the process ends at lower free energy. To use a familiar mechanical analogy, water at the top of a hill has the ability to perform the "work" of flowing downhill (i.e., it has high free energy), but, once it has flowed downhill, it cannot flow back up (i.e., it is in a state of low free energy). However, through another work process—that of a pump, for example—the water can be returned to the top of the hill, thereby recovering its ability to flow downhill. In thermodynamic terms, the free energy of the water has been increased by energy from an outside source (i.e., the pump). In the same way, the product molecules of a chemical reaction in a cell cannot reverse the reaction and return to their original state unless energy is supplied by coupling the process to another chemical reaction.

All catalysts, including enzymes, accelerate chemical reactions without affecting their direction. To return to the mechanical analogy, enzymes cannot make water flow uphill, although they can provide specific pathways for a downhill flow. Yet most of the chemical reactions that the cell needs to synthesize new molecules necessary for its growth require an uphill flow. In other words, the reactions require more energy than their starting molecules can provide.

Cells use a single strategy over and over again in order to get around the limitations of chemistry: they use the energy from an energy-releasing chemical reaction to drive an energy-absorbing reaction that would otherwise not occur. A useful mechanical analogy might be a mill wheel driven by the water in a stream. The water, in order to flow downhill, is forced to flow past the blades of the wheel, causing the wheel to turn. In this way, part of the energy from the moving stream is harnessed to move a mill wheel, which may be linked to a winch. As the winch turns, it can be used to pull a heavy load uphill. Thus, the energy-absorbing (but useful) uphill movement of a load can be driven by coupling it directly to the energy-releasing flow of water.

In cells, enzymes play the role of mill wheels by coupling energy-releasing reactions with energy-absorbing reactions. in cells the most important energy-releasing reaction serving a role similar to that of the flowing stream is the hydrolysis of adenosine triphosphate (ATP). In turn, the production of ATP molecules in the cells is an energy-absorbing reaction that is driven by being coupled to the energy-releasing breakdown of sugar molecules. In retracing this chain of reactions, it is necessary first to understand the source of the sugar molecules.

Photosynthesis: The beginning of the Food Chain

Sugar molecules are produced by the process of photosynthesis in plants and certain bacteria. These organisms lie at the base of the food chain, in that animals and other nonphotosynthesizing organisms depend on them for a constant supply of life-supporting organic molecules. Humans, for example, obtain these molecules by eating plants or other organisms that have previously eaten food derived from photosynthesizing organisms.

Plants and photosynthetic bacteria are unique in their ability to convert the freely available electromagnetic energy in sunlight into chemical bond energy, the energy that holds atoms together in molecules and is transferred or released in chemical reactions. The process of photosynthesis can be summarized by the following equation:

$$(\text{solar}) \text{ energy} + CO_2 + H_2O \rightarrow \text{sugar molecules} + O_2.$$

The energy-absorbing photosynthetic reaction is the reverse of the energy-releasing oxidative decomposition of sugar molecules. During photosynthesis, chlorophyll molecules absorb energy from sunlight and use it to fuel the production of simple sugars and other carbohydrates. The resulting abundance of sugar molecules and related biological products makes possible the existence of nonphotosynthesizing life on Earth.

ATP: Fueling Chemical Reactions

Certain enzymes catalyze the breakdown of organic foodstuffs. Once sugars are transported into cells, they either serve as building blocks in the form of amino acids for proteins and fatty acids for lipids or are subjected to metabolic pathways to provide the cell with ATP. ATP, the common carrier of energy inside the cell, is made from adenosine diphosphate (ADP) and inorganic phosphate (Pi). Stored in the chemical bond holding the terminal phosphate compound onto the ATP molecule is the energy derived from the breakdown of sugars. The removal of the terminal phosphate, through the water-mediated reaction called hydrolysis, releases this energy, which in turn fuels a large number of crucial energy-absorbing reactions in the cell. Hydrolysis can be summarized as follows:

$$ATP + H_2O \rightarrow ADP + P_i + \text{energy}.$$

The formation of ATP is the reverse of this equation, requiring the addition of energy. The central cellular pathway of ATP synthesis begins with glycolysis, a form of fermentation in which the sugar glucose is transformed into other sugars in a series of nine enzymatic reactions, each successive reaction involving an intermediate sugar containing phosphate. In the process, the six-carbon glucose is converted into two molecules of the three-carbon pyruvic acid. Some of the energy released through glycolysis of each glucose molecule is captured in the formation of two ATP molecules.

The second stage in the metabolism of sugars is a set of interrelated reactions called the tricarboxylic acid cycle. This cycle takes the three-carbon pyruvic acid produced in glycolysis and uses its carbon atoms to form carbon dioxide (CO_2) while transferring its hydrogen atoms to special carrier molecules, where they are held in high-energy linkage.

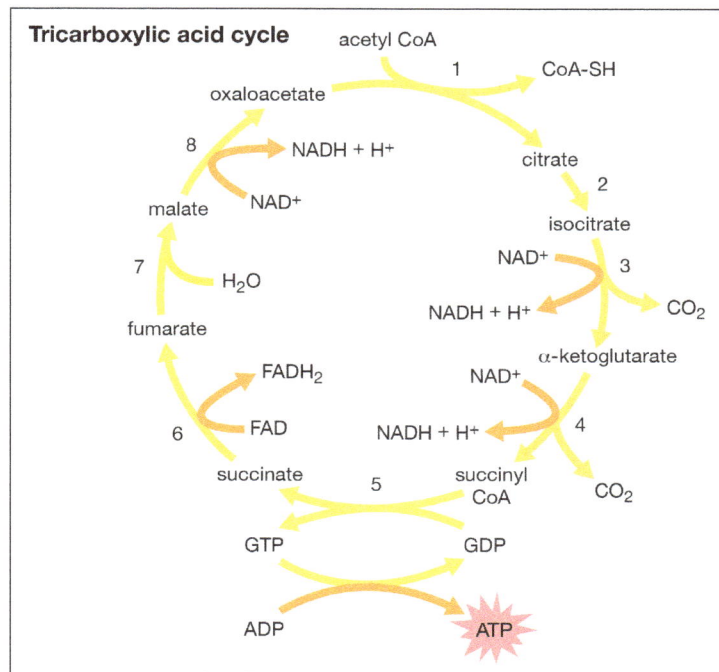

The eight-step tricarboxylic acid cycle.

In the third and last stage in the breakdown of sugars, oxidative phosphorylation, the high-energy hydrogen atoms are first separated into protons and high-energy electrons. The electrons are then passed from one electron carrier to another by means of an electron-transport chain. Each electron carrier in the chain has an increasing affinity for electrons, with the final electron acceptor being molecular oxygen (O_2). As separated electrons and protons, the hydrogen atoms are transferred to O_2 to form water. This reaction releases a large amount of energy, which drives the synthesis of a large number of ATP molecules from ADP and P_i.

Most of the cell's ATP is produced when the products of glycolysis are oxidized completely by a combination of the tricarboxylic acid cycle and oxidative phosphorylation. The process of glycolysis alone produces relatively small amounts of ATP. Glycolysis is an anaerobic reaction; that is, it can occur even in the absence of oxygen. The tricarboxylic acid cycle and oxidative phosphorylation, on the other hand, require oxygen. Glycolysis forms the basis of anaerobic fermentation, and it presumably was a major source of ATP for early life on Earth, when very little oxygen was available in the atmosphere. Eventually, however, bacteria evolved that were able to carry out photosynthesis. Photosynthesis liberated these bacteria from a dependence on the metabolism of organic materials that had accumulated from natural processes, and it also released oxygen into the atmosphere. Over a prolonged period of time, the concentration of molecular oxygen increased until it became freely available in the atmosphere. The aerobic tricarboxylic acid cycle and oxidative phosphorylation then evolved, and the resulting aerobic cells made much more efficient use of foodstuffs than their anaerobic ancestors, because they could convert much larger amounts of chemical bond energy into ATP.

The Genetic Information of Cells

Cells can thus be seen as a self-replicating network of catalytic macromolecules engaged in a carefully balanced series of energy conversions that drive biosynthesis and cell movement. But energy alone is not enough to make self-reproduction possible; the cell must contain detailed instructions that dictate exactly how that energy is to be used. These instructions are analogous to the blueprints that a builder uses to construct a house; in the case of cells, however, the blueprints themselves must be duplicated along with the cell before it divides, so that each daughter cell can retain the instructions that it needs for its own replication. These instructions constitute the cell's heredity.

DNA: The Genetic Material

During the early 19th century, it became widely accepted that all living organisms are composed of cells arising only from the growth and division of other cells. The improvement of the microscope then led to an era during which many biologists made intensive observations of the microscopic structure of cells. By 1885 a substantial amount of indirect evidence indicated that chromosomes—dark-staining threads in the cell nucleus—carried the information for cell heredity. It was later shown that chromosomes are about half DNA and half protein by weight.

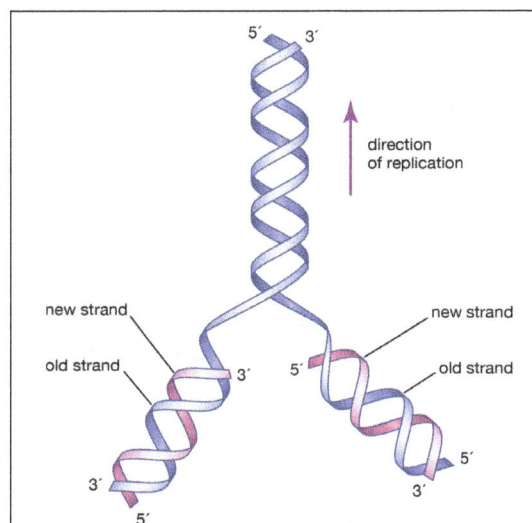

The initial proposal of the structure of DNA.

The revolutionary discovery suggesting that DNA molecules could provide the information for their own replication came in 1953, when American geneticist and biophysicist James Watson and British biophysicist Francis Crick proposed a model for the structure of the double-stranded DNA molecule (called the DNA double helix). In this model, each strand serves as a template in the synthesis of a complementary strand. Subsequent research confirmed the Watson and Crick model of DNA replication and showed that DNA carries the genetic information for reproduction of the entire cell.

All of the genetic information in a cell was initially thought to be confined to the DNA in the chromosomes of the cell nucleus. Later discoveries identified small amounts of additional genetic information present in the DNA of much smaller chromosomes located in two types of organelles in the cytoplasm. These organelles are the mitochondria in animal cells and the mitochondria and chloroplasts in plant cells. The special chromosomes carry the information coding for a few of the many proteins and RNA molecules needed by the organelles. They also hint at the evolutionary origin of these organelles, which are thought to have originated as free-living bacteria that were taken up by other organisms in the process of symbiosis.

RNA: Replicated from DNA

It is possible for RNA to replicate itself by mechanisms related to those used by DNA, even though it has a single-stranded instead of a double-stranded structure. In early cells RNA is thought to have replicated itself in this way. However, all of the RNA in present-day cells is synthesized by special enzymes that construct a single-stranded RNA chain by using one strand of the DNA helix as a template. Although RNA molecules are synthesized in the cell nucleus, where the DNA is located, most of them are transported to the cytoplasm before they carry out their functions.

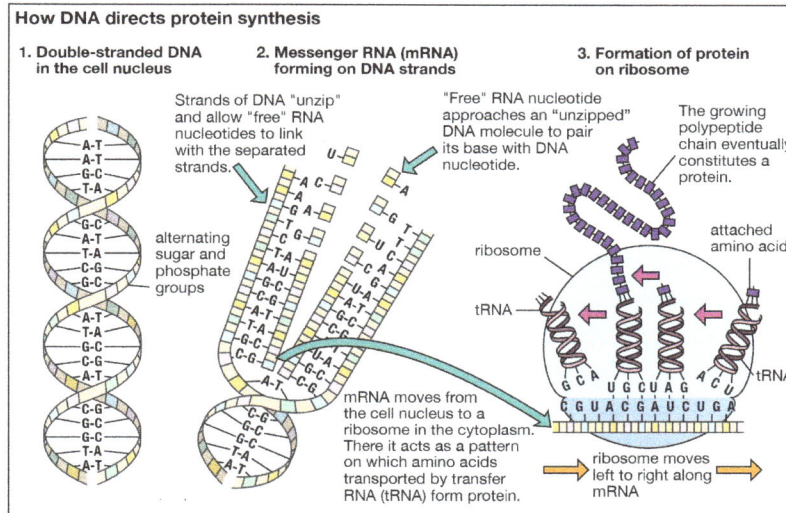

How DNA directs protein synthesis

Molecular genetics emerged from the realization that DNA and RNA constitute the genetic material of all living organisms. (1) DNA, located in the cell nucleus, is made up of nucleotides that contain the bases adenine (A), thymine (T), guanine (G), and cytosine (C). (2) RNA, which contains uracil (U) instead of thymine, transports the genetic code to protein-synthesizing sites in the cell. (3) Messenger RNA (mRNA) then carries the genetic information to ribosomes in the cell cytoplasm that translate the genetic information into molecules of protein.

The RNA molecules in cells have two main roles. Some, the ribozymes, fold up in ways that allow them to serve as catalysts for specific chemical reactions. Others serve as "messenger RNA," which provides templates specifying the synthesis of proteins. Ribosomes, tiny protein-synthesizing machines located in the cytoplasm, "read" the messenger RNA molecules and "translate" them into proteins by using the genetic code. In this translation, the sequence of nucleotides in the messenger RNA chain is decoded three nucleotides at a time, and each nucleotide triplet (called a codon) specifies a particular amino acid. Thus, a nucleotide sequence in the DNA specifies a protein provided that a messenger RNA molecule is produced from that DNA sequence. Each region of the DNA sequence specifying a protein in this way is called a gene.

DNA molecules catalyze not only their own duplication but also dictate the structures of all protein molecules. A single human cell contains about 10,000 different proteins produced by the expression of 10,000 different genes. Actually, a set of human chromosomes is thought to contain DNA with enough information to express between 30,000 and 100,000 proteins, but most of these proteins seem to be made only in specialized types of cells and are therefore not present throughout the body.

The Organization of Cells

Intracellular Communication

A cell with its many different DNA, RNA, and protein molecules is quite different from a test tube containing the same components. When a cell is dissolved in a test tube, thousands of different types of molecules randomly mix together. In the living cell, however, these components are kept in specific places, reflecting the high degree of organization essential for the growth and division of the cell. Maintaining this internal organization requires a continuous input of energy, because spontaneous chemical reactions always create disorganization. Thus, much of the energy released by ATP hydrolysis fuels processes that organize macromolecules inside the cell.

When a eukaryotic cell is examined at high magnification in an electron microscope, it becomes apparent that specific membrane-bound organelles divide the interior into a variety of subcompartments. Although not detectable in the electron microscope, it is clear from biochemical assays that each organelle contains a different set of macromolecules. This biochemical segregation reflects the functional specialization of each compartment. Thus, the mitochondria, which produce most of the cell's ATP, contain all of the enzymes needed to carry out the tricarboxylic acid cycle and oxidative phosphorylation. Similarly, the degradative enzymes needed for the intracellular digestion of unwanted macromolecules are confined to the lysosomes.

The relative volumes occupied by some cellular compartments in a typical liver cell		
cellular compartment	percent of total cell volume	approximate number per cell
cytosol	54	1
mitochondrion	22	1,700
endoplasmic reticulum plus Golgi apparatus	15	1
nucleus	6	1
lysosome	1	300

It is clear from this functional segregation that the many different proteins specified by the genes in the cell nucleus must be transported to the compartment where they will be used. Not surprisingly, the cell contains an extensive membrane-bound system devoted to maintaining just this intracellular order. The system serves as a post office, guaranteeing the proper routing of newly synthesized macromolecules to their proper destinations.

All proteins are synthesized on ribosomes located in the cytosol. As soon as the first portion of the amino acid sequence of a protein emerges from the ribosome, it is inspected for the presence of a short "endoplasmic reticulum (ER) signal sequence." Those ribosomes making proteins with such a sequence are transported to the surface of the ER membrane, where they complete their synthesis; the proteins made on these ribosomes are immediately transferred through the ER membrane to the inside of the ER compartment. Proteins lacking the ER signal sequence remain in the cytosol and are released from the ribosomes when their synthesis is completed. This chemical decision process places some newly completed protein chains in the cytosol and others within an extensive membrane-bounded compartment in the cytoplasm, representing the first step in intracellular protein sorting.

The newly made proteins in both cell compartments are then sorted further according to additional signal sequences that they contain. Some of the proteins in the cytosol remain there, while others go to the surface of mitochondria or (in plant cells) chloroplasts, where they are transferred through the membranes into the organelles. Subsignals on each of these proteins then designate exactly where in the organelle the protein belongs. The proteins initially sorted into the ER have an even wider range of destinations. Some of them remain in the ER, where they function as part of the organelle. Most enter transport vesicles and pass to the Golgi apparatus, separate membrane-bounded organelles that contain at least three subcompartments. Some of the proteins are retained in the subcompartments of the Golgi, where they are utilized for functions peculiar to that organelle. Most eventually enter vesicles that leave the Golgi for other cellular destinations such as the cell membrane, lysosomes, or special secretory vesicles.

Intercellular Communication

Formation of a multicellular organism starts with a small collection of similar cells in an embryo and proceeds by continuous cell division and specialization to produce an entire community of cooperating cells, each with its own role in the life of the organism. Through cell cooperation, the organism becomes much more than the sum of its component parts.

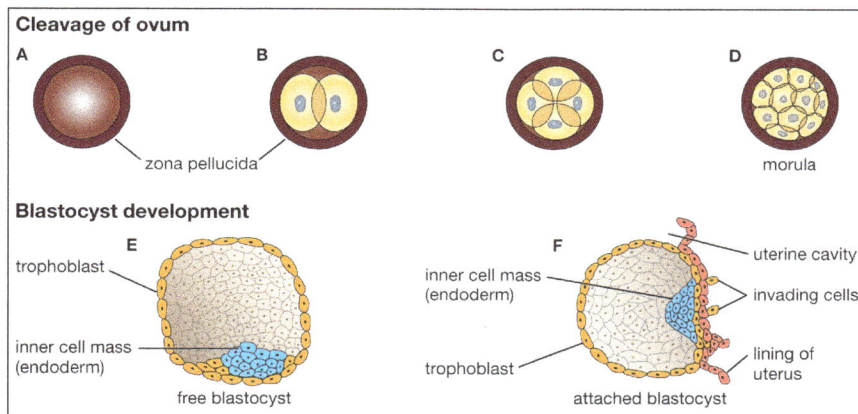

The ovum contains a small collection of cells in the early stages of human development.

As cells divide (A–D), they are separated into different regions of the ovum. Each region of the ovum transmits a unique set of chemical signals to nearby cells. Thus, the signals detected by one cell differ from those detected by its neighbour cells. In this process, known as cell determination, cells are individually programmed to direct them toward development into different cell types.

A fertilized egg multiplies and produces a whole family of daughter cells, each of which adopts a structure and function according to its position in the entire assembly. All of the daughter cells contain the same chromosomes and therefore the same genetic information. Despite this common inheritance, different types of cells behave differently and have different structures. In order for this to be the case, they must express different sets of genes, so that they produce different proteins despite their identical embryological ancestors.

During the development of an embryo, it is not sufficient for all the cell types found in the fully developed individual simply to be created. Each cell type must form in the right place at the right time and in the correct proportion; otherwise, there would be a jumble of randomly assorted cells in no way resembling an organism. The orderly development of an organism depends on a process called cell determination, in which initially identical cells become committed to different pathways of development. A fundamental part of cell determination is the ability of cells to detect different chemicals within different regions of the embryo. The chemical signals detected by one cell may be different from the signals detected by its neighbour cells. The signals that a cell detects activate a set of genes that tell the cell to differentiate in ways appropriate for its position within the embryo. The set of genes activated in one cell differs from the set of genes activated in the cells around it. The process of cell determination requires an elaborate system of cell-to-cell communication in early embryos.

The Cell Membrane

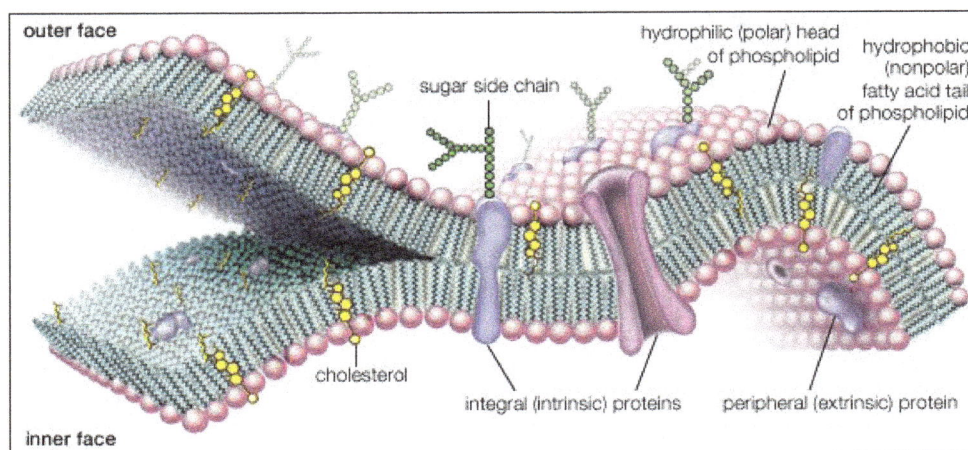

Molecular view of the cell membrane.

Intrinsic proteins penetrate and bind tightly to the lipid bilayer, which is made up largely of phospholipids and cholesterol and which typically is between 4 and 10 nanometers (nm; 1 nm = 10−9 metre) in thickness. Extrinsic proteins are loosely bound to the hydrophilic (polar) surfaces, which face the watery medium both inside and outside the cell. Some intrinsic proteins present sugar side chains on the cell's outer surface.

A thin membrane, typically between 4 and 10 nanometers (nm; 1 nm = 10−9 metre) in thickness, surrounds every living cell, delimiting the cell from the environment around it. Enclosed by this cell membrane (also known as the plasma membrane) are the cell's constituents, often large, water-soluble, highly charged molecules such as proteins, nucleic acids, carbohydrates, and substances involved in cellular metabolism. Outside the cell, in the surrounding water-based environment, are ions, acids, and alkalis that are toxic to the cell, as well as nutrients that the cell must absorb in order to live and grow. The cell membrane, therefore, has two functions: first, to be a barrier keeping the constituents of the cell in and unwanted substances out and, second, to be a gate allowing transport into the cell of essential nutrients and movement from the cell of waste products.

Chemical Composition and Membrane Structure

Most current knowledge about the biochemical constituents of cell membranes originates in studies of red blood cells. The chief advantage of these cells for experimental purposes is that they may be obtained easily in large amounts and that they have no internal membranous organelles to interfere with study of their cell membranes. Careful studies of these and other cell types have shown that all membranes are composed of proteins and fatty-acid-based lipids. Membranes actively involved in metabolism contain a higher proportion of protein; thus, the membrane of the mitochondrion, the most rapidly metabolizing organelle of the cell, contains as much as 75 percent protein, while the membrane of the Schwann cell, which forms an insulating sheath around many nerve cells, has as little as 20 percent protein.

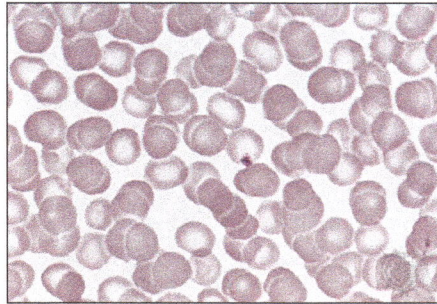

Human red blood cells (erythrocytes).

Membrane Lipids

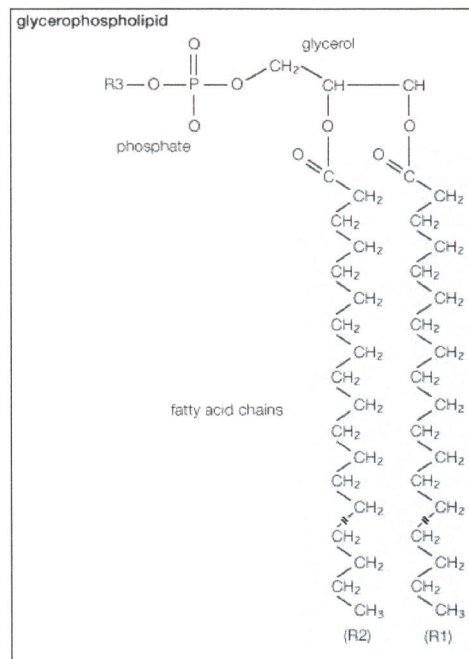

General structural formula of a glycerophospholipid. The composition of the specific molecule depends on the chemical group (designated R3 in the diagram) linked to the phosphate and glycerol "head" and also on the lengths of the fatty acid "tails" (R1 and R2).

Membrane lipids are principally of two types, phospholipids and sterols (generally cholesterol). Both types share the defining characteristic of lipids—they dissolve readily in organic solvents—but in addition they both have a region that is attracted to and soluble in water. This "amphiphilic" property (having a dual attraction; i.e., containing both a lipid-soluble and a water-soluble region) is basic to the role of lipids as building blocks of cellular membranes. Phospholipid molecules have a head (often of glycerol) to which are attached two long fatty acid chains that look much like tails. These tails are repelled by water and dissolve readily in organic solvents, giving the molecule its lipid character. To another part of the head is attached a phosphoryl group with a negative electrical charge; to this group in turn is attached another group with a positive or neutral charge. This portion of the phospholipid dissolves in water, thereby completing the molecule's amphiphilic character. In contrast, sterols have a complex hydrocarbon ring structure as the lipid-soluble region and a hydroxyl grouping as the water-soluble region.

When dry phospholipids, or a mixture of such phospholipids and cholesterol, are immersed in water under laboratory conditions, they spontaneously form globular structures called liposomes. Investigation of the liposomes shows them to be made of concentric spheres, one sphere inside of another and each forming half of a bilayered wall. A bilayer is composed of two sheets of phospholipid molecules with all of the molecules of each sheet aligned in the same direction. In a water medium, the phospholipids of the two sheets align so that their water-repellent, lipid-soluble tails are turned and loosely bonded to the tails of the molecules on the other sheet. The water-soluble heads turn outward into the water, to which they are chemically attracted. In this way, the two sheets form a fluid, sandwichlike structure, with the fatty acid chains in the middle mingling in an organic medium while sealing out the water medium.

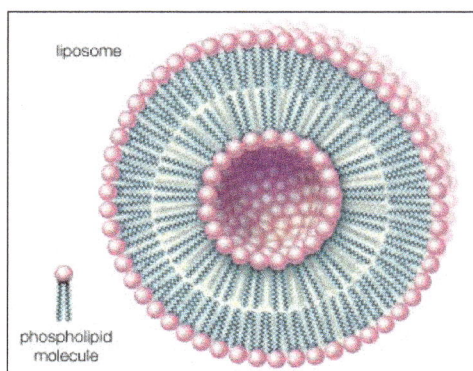

Phospholipids can be used to form artificial structures called liposomes, which are double-walled, hollow spheres useful for encapsulating other molecules such as pharmaceutical drugs.

This type of lipid bilayer, formed by the self-assembly of lipid molecules, is the basic structure of the cell membrane. It is the most stable thermodynamic structure that a phospholipid-water mixture can take up: the fatty acid portion of each molecule dissolved in the organic phase formed by the identical regions of the other molecules and the water-attractive regions surrounded by water and facing away from the fatty acid regions. The chemical affinity of each region of the amphiphilic molecule is thus satisfied in the bilayer structure.

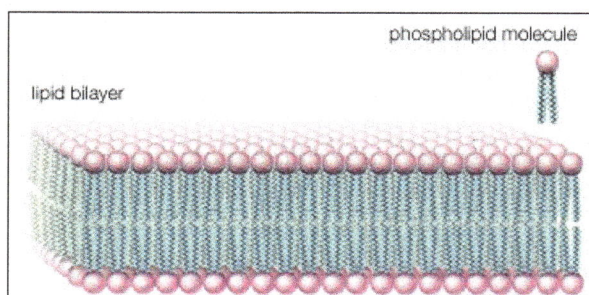

Phospholipid molecules, like molecules of many lipids, are composed of a hydrophilic "head" and one or more hydrophobic "tails." In a water medium, the molecules form a lipid bilayer, or two-layered sheet, in which the heads are turned toward the watery medium and the tails are sheltered inside, away from the water. This bilayer is the basis of the membranes of living cells.

Membrane Proteins

Membrane proteins are also of two general types. One type, called the extrinsic proteins, is loosely attached by ionic bonds or calcium bridges to the electrically charged phosphoryl surface of the bilayer. They can also attach to the second type of protein, called the intrinsic proteins. The intrinsic proteins, as their name implies, are firmly embedded within the phospholipid bilayer. Almost all intrinsic proteins contain special amino acid sequences, generally about 20- to 24-amino acids long, that extend through the internal regions of the cell membrane.

Most intrinsic and extrinsic proteins bear on their outer surfaces side chains of complex sugars, which extend into the aqueous environment around the cell. For this reason, these proteins are often referred to as glycoproteins. Some glycoproteins are involved in cell-to-cell recognition.

Membrane Fluidity

One of the triumphs of cell biology during the decade from 1965 to 1975 was the recognition of the cell membrane as a fluid collection of amphiphilic molecules. This array of proteins, sterols, and phospholipids is organized into a liquid crystal, a structure that lends itself readily to rapid cell growth. Measurements of the membrane's viscosity show it as a fluid one hundred times as viscous as water, similar to a thin oil. The phospholipid molecules diffuse readily in the plane of the bilayer. Many of the membrane's proteins also have this freedom of movement, but some are fixed in the membrane by interaction with the cell's cytoskeleton. Newly synthesized phospholipids insert themselves easily into the existing cell membrane. Intrinsic proteins are inserted during their synthesis on ribosomes bound to the endoplasmic reticulum, whereas extrinsic proteins found on the internal surface of the cell membrane are synthesized on free, or unattached, ribosomes, liberated into the cytoplasm, and then brought to the membrane.

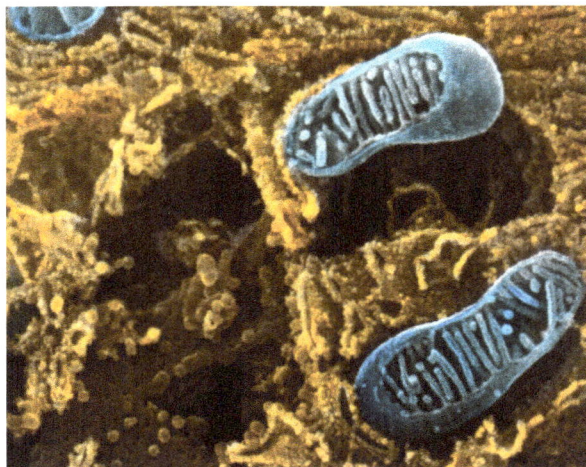

Endoplasmic reticulum; organelle. A scanning electron micrograph of a pancreatic acinar cell, showing mitochondria (blue), rough endoplasmic reticulum (yellow; ribosomes appear as small dots), and Golgi apparatus (gray, at centre and lower left).

Internal Membranes

The presence of internal membranes distinguishes eukaryotic cells (cells with a nucleus) from prokaryotic cells (those without a nucleus). Prokaryotic cells are small (one to five micrometres in length) and contain only a single cell membrane; metabolic functions are often confined to different patches of the membrane rather than to areas in the body of the cell. Typical eukaryotic cells, by contrast, are much larger, the cell membrane constituting only 10 percent or less of the total cellular membrane. Metabolic functions in these cells are carried out in the organelles, compartments sequestered from the cell body, or cytoplasm, by internal membranes.

The lysosomes, peroxisomes, and (in plants) glyoxysomes enclose extremely reactive by-products and enzymes. Internal membranes form the mazelike endoplasmic reticulum, where cell membrane proteins and lipids are synthesized, and they also form the stacks of flattened sacs called the Golgi apparatus, which is associated with the transport and modification of lipids, proteins, and carbohydrates. Finally, internal cell membranes can form storage and transport vesicles and the vacuoles of plant cells. Each membrane structure has its own distinct composition of proteins and lipids enabling it to carry out unique functions.

The Mitochondrion and the Chloroplast

Mitochondria and chloroplasts are the powerhouses of the cell. Mitochondria appear in both plant and animal cells as elongated cylindrical bodies, roughly one micrometre in length and closely packed in regions actively using metabolic energy. Mitochondria oxidize the products of cytoplasmic metabolism to generate adenosine triphosphate (ATP), the energy currency of the cell. Chloroplasts are the photosynthetic organelles in plants and some algae. They trap light energy and convert it partly into ATP but mainly into certain chemically reduced molecules that, together with ATP, are used in the first steps of carbohydrate production. Mitochondria and chloroplasts share a certain structural resemblance, and both have a somewhat independent existence within the cell, synthesizing some proteins from instructions supplied by their own DNA.

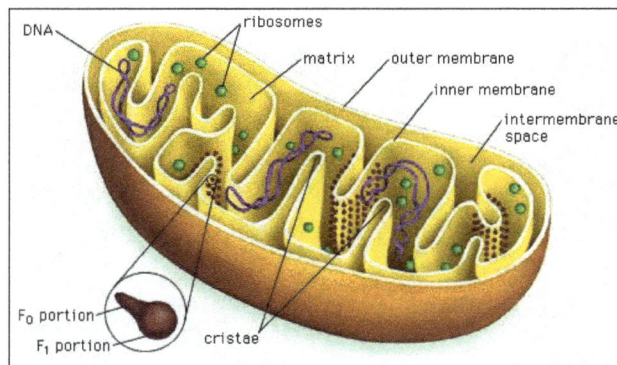

The internal membrance of a mitochondrion is elaborately folded into structures known as cristae. Cristae increase the surface area of the inner membrane, which houses the components of the electron-transport chain. Protein known as F_1F_0ATPases that produce the majority of ATP used by cells are found throughout the cristae.

Mitochondrial and Chloroplastic Structure

Both organelles are bounded by an external membrane that serves as a barrier by blocking the passage of cytoplasmic proteins into the organelle. An inner membrane provides an additional barrier that is impermeable even to small ions such as protons. The membranes of both organelles have a lipid bilayer construction. Located between the inner and outer membranes is the inter-membrane space.

In mitochondria the inner membrane is elaborately folded into structures called cristae that dramatically increase the surface area of the membrane. In contrast, the inner membrane of chloroplasts is relatively smooth. However, within this membrane is yet another series of folded membranes that form a set of flattened, disklike sacs called thylakoids. The space enclosed by the inner membrane is called the matrix in mitochondria and the stroma in chloroplasts. Both spaces are filled with a fluid containing a rich mixture of metabolic products, enzymes, and ions. Enclosed by the thylakoid membrane of the chloroplast is the thylakoid space. The extraordinary chemical capabilities of the two organelles lie in the cristae and the thylakoids. Both membranes are studded with enzymatic proteins either traversing the bilayer or dissolved within the bilayer. These proteins contribute to the production of energy by transporting material across the membranes and by serving as electron carriers in important oxidation-reduction reactions.

Metabolic Functions

Crucial to the function of mitochondria and chloroplasts is the chemistry of the oxidation-reduction, or redox, reaction. This controlled burning of material comprises the transfer of electrons from one compound, called the donor, to another, called the acceptor. All compounds taking part in redox reactions are ranked in a descending scale according to their ability to act as electron donors. Those higher in the scale donate electrons to their fellows lower down, which have a lesser tendency to donate, but a correspondingly greater tendency to accept, electrons. Each acceptor in turn donates electrons to the next compound down the scale, forming a donor-acceptor chain extending from the greatest donating ability to the least.

At the top of the scale is hydrogen, the most abundant element in the universe. The nucleus of a hydrogen atom is composed of one positively charged proton; around the nucleus revolves one negatively charged electron. In the atmosphere two hydrogen atoms join to form a hydrogen molecule (H_2). In solution the two atoms pull apart, dissociating into their constituent protons and electrons. In the redox reaction the electrons are passed from one reactant to another. The donation of electrons is called oxidation, and the acceptance is called reduction—hence the descriptive term oxidation-reduction, indicating that one action never takes place without the other.

A hydrogen atom has a great tendency to transfer an electron to an acceptor. An oxygen atom, in contrast, has a great tendency to accept an electron. The burning of hydrogen by oxygen is, chemically, the transfer of an electron from each of two hydrogen atoms to oxygen, so that hydrogen is oxidized and oxygen reduced. The reaction is extremely exergonic; i.e., it liberates much free energy as heat. This is the reaction that takes place within mitochondria but is so controlled that the heat is liberated not at once but in a series of steps. The free energy, harnessed by the organelle, is coupled to the synthesis of ATP from adenosine diphosphate (ADP) and inorganic phosphate (P_i).

An analogy can be drawn between this controlled reaction and the flow of river water down a lock system. Without the locks, water flow would be rapid and uncontrolled, and no ship could safely ply the river. The locks force water to flow in small controlled steps conducive to safe navigation.

But there is more to a lock system than this. The flow of water down the locks can also be harnessed to raise a ship from a lower to a higher level, with the water rather than the ship expending the energy. In mitochondria the burning of hydrogen is broken into a series of small indirect steps following the flow of electrons along a chain of donor-acceptors. Energy is funneled into the chemical bonding of ADP and Pi, raising the free energy of these two compounds to the high level of ATP.

Cell-to-cell Communication via Chemical Signaling

In addition to cell-matrix and cell-cell interactions, cell behaviour in multicellular organisms is coordinated by the passage of chemical or electrical signals between cells. The most common form of chemical signaling is via molecules secreted from the cells and moving through the extracellular space. Signaling molecules may also remain on cell surfaces, influencing other cells only after the cells make physical contact. Finally, gap junctions allow small molecules to move between the cytoplasms of adjacent cells.

Types of Chemical Signaling

Chemical signals secreted by cells can act over varying distances. In the autocrine signaling process, molecules act on the same cells that produce them. In paracrine signaling, they act on nearby cells. Autocrine signals include extracellular matrix molecules and various factors that stimulate cell growth. An example of paracrine signals is the chemical transmitted from nerve to muscle that causes the muscle to contract. In this instance, the muscle cells have regions specialized to receive chemical signals from an adjacent nerve cell. In both autocrine and paracrine signaling, the chemical signal works in the immediate vicinity of the cell that produces it and is present at high concentrations. A chemical signal picked up by the bloodstream and taken to distant sites is called an endocrine signal. Most hormones produced in vertebrates are endocrine signals, such as the hormones produced in the pituitary gland at the base of the brain and carried by the bloodstream to act at low concentrations on the thyroid or adrenal glands.

The concentration at which a chemical signal acts has significance for its target cell. Chemical signals that act at high concentration act locally and rapidly. On the other hand, chemical signals that act at low concentrations act at distances and are generally slow.

Signal Receptors

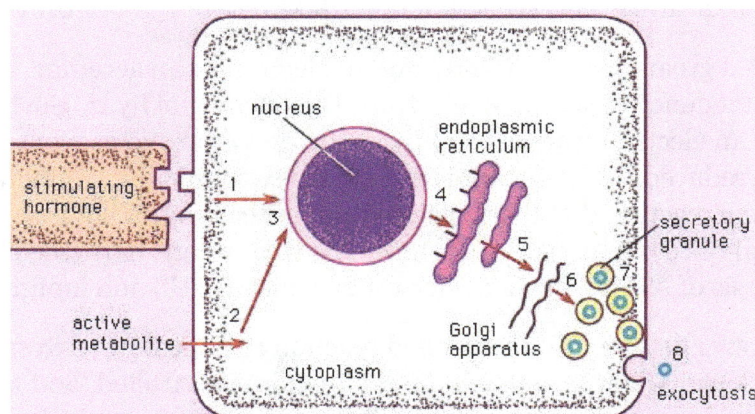

The ability of a cell to respond to an extracellular signal depends on the presence of specific proteins called receptors, which are located on the cell surface or in the cytoplasm. Receptors bind chemical signals that ultimately trigger a mechanism to modify the behaviour of the target cell. Cells may contain an array of specific receptors that allow them to respond to a variety of chemical signals.

Hormones and active metabolites bind to different types of receptors. Water-soluble molecules (i.e., insulin) cannot pass through the lipid membrane of a cell and thus rely on cell surface receptors to transmit messages to the interior of the cell. In contrast, lipid-soluble molecules (i.e., certain active metabolites) are able to diffuse through the lipid membrane to communicate messages directly to the nucleus.

Signal molecules are either soluble or insoluble. Water-soluble molecules, such as the polypeptide hormone insulin, bind to receptors at cell surfaces. On the other hand, lipid-soluble molecules, such as the steroid hormones produced by the ovary or testis, pass through the lipid bilayer of the cell membrane to reach receptors within the cytoplasm. Extracellular matrix molecules are chemical signals, but, because of their size and insolubility, they act only on cell surface receptors and are neither taken up by the cells nor rapidly destroyed.

Cellular Response

The binding of chemical signals to their corresponding receptors induces events within the cell that ultimately change its behaviour. The nature of these intracellular events differs according to the type of receptor. Also, the same chemical signal can trigger different responses in different types of cell.

Cell surface receptors work in several ways when they are occupied. Some receptors enter the cell still bound to the chemical signal. Others activate membrane enzymes, which produce certain intracellular chemical mediators. Still other receptors open membrane channels, allowing a flow of ions that causes either a change in the electrical properties of the membrane or a change in the ion concentration in the cytoplasm. This regulation of enzymes or membrane channels produces changes in the concentration of intracellular signaling molecules, which are often called second messengers (the first messenger being the extracellular chemical signal bound to the receptor).

Two common intracellular signaling molecules are cyclic AMP and the calcium ion. Cyclic AMP is a derivative of adenosine triphosphate, the ubiquitous energy-carrying molecule of the cell. The intracellular concentrations of both cyclic AMP and calcium ions are normally very low. The binding of an extracellular chemical signal to a cell surface receptor stimulates an enzyme complex in the membrane to produce cyclic AMP. This second messenger then diffuses into the cytoplasm and acts on intracellular enzymes called kinases that modify the behaviour of the cell, culminating in the activation of target genes that increase the synthesis of certain proteins. The action of cyclic AMP is brief because it is rapidly degraded by specific enzymes.

Occupancy of other surface receptors causes a transient opening of membrane channels. This can allow calcium ions to enter the cytoplasm from the extracellular space, where their concentration is higher. The action of calcium ions is also brief because they are rapidly pumped out of the cell or bound to intracellular molecules, lowering the cytoplasmic concentration to the state existing before stimulation.

Some extracellular chemical signals enter the cell intact, still bound to the receptor, without generating a second messenger. In this mechanism, receptor occupancy causes individual receptors within the cell membrane to aggregate spontaneously. That portion of the membrane containing the aggregated receptors is then taken into the cell, where it fuses with various membrane-bounded organelles in the cytoplasm. In some instances the chemical signal is released within the organelles, and in almost all instances the ingested membrane is rapidly returned to the cell membrane along with the surface receptors.

The Plant Cell Wall

The plant cell wall is a specialized form of extracellular matrix that surrounds every cell of a plant and is responsible for many of the characteristics distinguishing plant from animal cells. Although often perceived as an inactive product serving mainly mechanical and structural purposes, the cell wall actually has a multitude of functions upon which plant life depends. Such functions include: (1) providing the protoplast, or living cell, with mechanical protection and a chemically buffered environment, (2) providing a porous medium for the circulation and distribution of water, minerals, and other small nutrient molecules, (3) providing rigid building blocks from which stable structures of higher order, such as leaves and stems, can be produced, and (4) providing a storage site of regulatory molecules that sense the presence of pathogenic microbes and control the development of tissues.

Mechanical Properties of Wall Layers

All cell walls contain two layers, the middle lamella and the primary cell wall, and many cells produce an additional layer, called the secondary wall. The middle lamella serves as a cementing layer between the primary walls of adjacent cells. The primary wall is the cellulose-containing layer laid down by cells that are dividing and growing. To allow for cell wall expansion during growth, primary walls are thinner and less rigid than those of cells that have stopped growing. A fully grown plant cell may retain its primary cell wall (sometimes thickening it), or it may deposit an additional, rigidifying layer of different composition; this is the secondary wall. Secondary cell walls are responsible for most of the plant's mechanical support as well as the mechanical properties prized in wood. In contrast to the permanent stiffness and load-bearing capacity of thick secondary walls, the thin primary walls are capable of serving a structural, supportive role only when the vacuoles within the cell are filled with water to the point that they exert a turgor pressure against the cell wall. Turgor-induced stiffening of primary walls is analogous to the stiffening of the sides of a pneumatic tire by air pressure. The wilting of flowers and leaves is caused by a loss of turgor pressure, which results in turn from the loss of water from the plant cells.

Components of the Cell Wall

Although primary and secondary wall layers differ in detailed chemical composition and structural organization, their basic architecture is the same, consisting of cellulose fibres of great tensile strength embedded in a water-saturated matrix of polysaccharides and structural glycoproteins.

Cellulose

Cellulose consists of several thousand glucose molecules linked end to end. The chemical links between the individual glucose subunits give each cellulose molecule a flat ribbonlike structure that allows adjacent molecules to band laterally together into microfibrils with lengths ranging from two to seven micrometres. Cellulose fibrils are synthesized by enzymes floating in the cell membrane and are arranged in a rosette configuration. Each rosette appears capable of "spinning" a microfibril into the cell wall. During this process, as new glucose subunits are added to the growing end of the fibril, the rosette is pushed around the cell on the surface of the cell membrane, and its cellulose fibril becomes wrapped around the protoplast. Thus, each plant cell can be viewed as making its own cellulose fibril cocoon.

Matrix Polysaccharides

The two major classes of cell wall matrix polysaccharides are the hemicelluloses and the pectic polysaccharides, or pectins. Both are synthesized in the Golgi apparatus, brought to the cell surface in small vesicles, and secreted into the cell wall.

Hemicelluloses consist of glucose molecules arranged end to end as in cellulose, with short side chains of xylose and other uncharged sugars attached to one side of the ribbon. The other side of the ribbon binds tightly to the surface of cellulose fibrils, thereby coating the microfibrils with hemicellulose and preventing them from adhering together in an uncontrolled manner. Hemicellulose molecules have been shown to regulate the rate at which primary cell walls expand during growth.

The heterogeneous, branched, and highly hydrated pectic polysaccharides differ from hemicelluloses in important respects. Most notably, they are negatively charged because of galacturonic acid residues, which, together with rhamnose sugar molecules, form the linear backbone of all pectic polysaccharides. The backbone contains stretches of pure galacturonic acid residues interrupted by segments in which galacturonic acid and rhamnose residues alternate; attached to these latter segments are complex, branched sugar side chains. Because of their negative charge, pectic polysaccharides bind tightly to positively charged ions, or cations. In cell walls, calcium ions cross-link the stretches of pure galacturonic acid residues tightly, while leaving the rhamnose-containing segments in a more open, porous configuration. This cross-linking creates the semirigid gel properties characteristic of the cell wall matrix—a process exploited in the preparation of jellied preserves.

Proteins

Although plant cell walls contain only small amounts of protein, they serve a number of important functions. The most prominent group are the hydroxyproline-rich glycoproteins, shaped like rods with connector sites, of which extensin is a prominent example. Extensin contains 45 percent hydroxyproline and 14 percent serine residues distributed along its length. Every hydroxyproline residue carries a short side chain of arabinose sugars, and most serine residues carry a galactose sugar; this gives rise to long molecules, resembling bottle brushes, that are secreted into the cell wall toward the end of primary-wall formation and become covalently cross-linked into a mesh at the time that cell growth stops. Plant cells may control their ultimate size by regulating the time at which this cross-linking of extensin molecules occurs.

In addition to the structural proteins, cell walls contain a variety of enzymes. Most notable are those that cross-link extensin, lignin, cutin, and suberin molecules into networks. Other enzymes help protect plants against fungal pathogens by breaking fragments off of the cell walls of the fungi. The fragments in turn induce defense responses in underlying cells. The softening of ripe fruit and dropping of leaves in the autumn are brought about by cell wall-degrading enzymes.

Plastics

Cell wall plastics such as lignin, cutin, and suberin all contain a variety of organic compounds cross-linked into tight three-dimensional networks that strengthen cell walls and make them more resistant to fungal and bacterial attack. Lignin is the general name for a diverse group of polymers of aromatic alcohols. Deposited mostly in secondary cell walls and providing the rigidity of terrestrial vascular plants, it accounts for up to 30 percent of a plant's dry weight. The diversity of cross-links between the polymers—and the resulting tightness—makes lignin a formidable barrier to the penetration of most microbes.

Agave shawii (top) and Echeveria (bottom), two types of xerophytes (plants adapted to arid habitats). They develop highly cutinized fleshy leaves and stems for water storage with which they modulate the effects of strong sunlight, low humidity, and scant water.

Cutin and suberin are complex biopolyesters composed of fatty acids and aromatic compounds. Cutin is the major component of the cuticle, the waxy, water-repelling surface layer of cell walls exposed to the environment aboveground. By reducing the wetability of leaves and stems—and thereby affecting the ability of fungal spores to germinate—it plays an important part in the defense strategy of plants. Suberin serves with waxes as a surface barrier of underground parts. Its synthesis is also stimulated in cells close to wounds, thereby sealing off the wound surfaces and protecting underlying cells from dehydration.

Intercellular Communication

Plasmodesmata

Similar to the gap junction of animal cells is the plasmodesma, a channel passing through the cell wall and allowing direct molecular communication between adjacent plant cells. Plasmodesmata are lined with cell membrane, in effect uniting all connected cells with one continuous cell membrane. Running down the middle of each channel is a thin membranous tube that connects the endoplasmic reticula (ER) of the two cells. This structure is a remnant of the ER of the original parent cell, which, as the parent cell divided, was caught in the developing cell plate.

Although the precise mechanisms are not fully understood, the plasmodesma is thought to regulate the passage of small molecules such as salts, sugars, and amino acids by constricting or dilating the openings at each end of the channel.

Oligosaccharides with Regulatory Functions

The discovery of cell wall fragments with regulatory functions opened a new era in plant research. For years scientists had been puzzled by the chemical complexity of cell wall polysaccharides, which far exceeds the structural requirements of plant cell walls. The answer came when it was found that specific fragments of cell wall polysaccharides, called oligosaccharins, are able to induce specific responses in plant cells and tissues. One such fragment, released by enzymes used by fungi to break down plant cell walls, consists of a linear polymer of 10 to 12 galacturonic acid residues. Exposure of plant cells to such fragments induces them to produce antibiotics known as phytoalexins. In other experiments it has been shown that exposing strips of tobacco stem cells to a different type of cell wall fragment leads to the growth of roots; other fragments lead to the formation of stems, and yet others to the production of flowers. In all instances the concentration of oligosaccharins required to bring about the observed responses is equal to that of hormones in animal cells; indeed, oligosaccharins may be viewed as the oligosaccharide hormones of plants.

Cell Division and Growth

In unicellular organisms, cell division is the means of reproduction; in multicellular organisms, it is the means of tissue growth and maintenance. Survival of the eukaryotes depends upon interactions between many cell types, and it is essential that a balanced distribution of types be maintained. This is achieved by the highly regulated process of cell proliferation. The growth and division of different cell populations are regulated in different ways, but the basic mechanisms are similar throughout multicellular organisms.

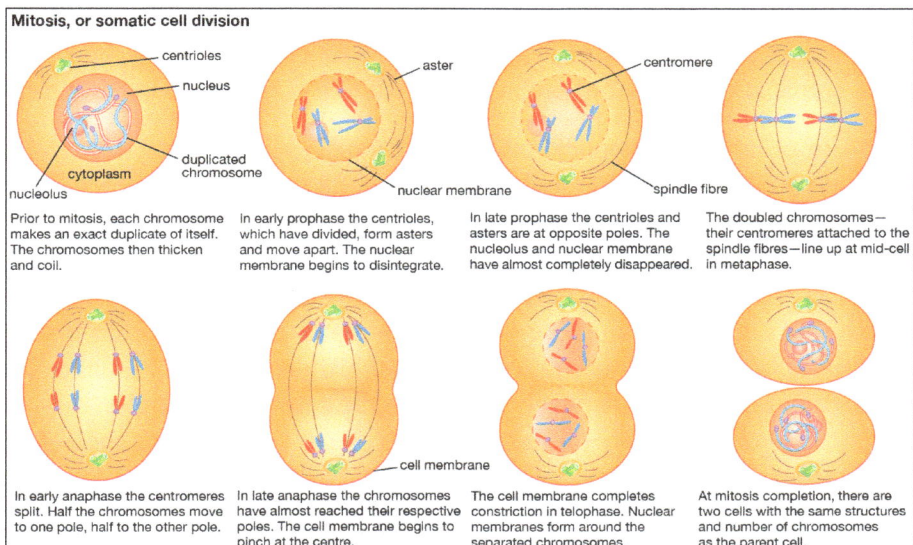

One cell gives rise to two genetically identical daughter cells during the process of mitosis.

Most tissues of the body grow by increasing their cell number, but this growth is highly regulated to maintain a balance between different tissues. In adults most cell division is involved in tissue renewal rather than growth, many types of cells undergoing continuous replacement. Skin cells, for example, are constantly being sloughed off and replaced; in this case, the mature differentiated cells do not divide, but their population is renewed by division of immature stem cells. In certain other cells, such as those of the liver, mature cells remain capable of division to allow growth or regeneration after injury.

In contrast to these patterns, other types of cells either cannot divide or are prevented from dividing by certain molecules produced by nearby cells. As a result, in the adult organism, some tissues have a greatly reduced capacity to renew damaged or diseased cells. Examples of such tissues include heart muscle, nerve cells of the central nervous system, and lens cells in mammals. Maintenance and repair of these cells is limited to replacing intracellular components rather than replacing entire cells.

Duplication of the Genetic Material

Before a cell can divide, it must accurately and completely duplicate the genetic information encoded in its DNA in order for its progeny cells to function and survive. This is a complex problem because of the great length of DNA molecules. Each human chromosome consists of a long double spiral, or helix, each strand of which consists of more than 100 million nucleotides.

The duplication of DNA is called DNA replication, and it is initiated by complex enzymes called DNA polymerases. These progress along the molecule, reading the sequences of nucleotides that are linked together to make DNA chains. Each strand of the DNA double helix, therefore, acts as a template specifying the nucleotide structure of a new growing chain. After replication, each of the two daughter DNA double helices consists of one parental DNA strand wound around one newly synthesized DNA strand.

In order for DNA to replicate, the two strands must be unwound from each other. Enzymes called helicases unwind the two DNA strands, and additional proteins bind to the separated strands to stabilize them and prevent them from pairing again. In addition, a remarkable class of enzyme called DNA topoisomerase removes the helical twists by cutting either one or both strands and then resealing the cut. These enzymes can also untangle and unknot DNA when it is tightly coiled into a chromatin fibre.

In the circular DNA of prokaryotes, replication starts at a unique site called the origin of replication and then proceeds in both directions around the molecule until the two processes meet, producing two daughter molecules. In rapidly growing prokaryotes, a second round of replication can start before the first has finished. The situation in eukaryotes is more complicated, as replication moves more slowly than in prokaryotes. At 500 to 5,000 nucleotides per minute (versus 100,000 nucleotides per minute in prokaryotes), it would take a human chromosome about a month to replicate if started at a single site. Actually, replication begins at many sites on the long chromosomes of animals, plants, and fungi. Distances between adjacent initiation sites are not always the same; for example, they are closer in the rapidly dividing embryonic cells of frogs or flies than in adult cells of the same species.

Accurate DNA replication is crucial to ensure that daughter cells have exact copies of the genetic information for synthesizing proteins. Accuracy is achieved by a "proofreading" ability of the DNA polymerase itself. It can erase its own errors and then synthesize anew. There are also repair systems that correct genetic damage to DNA. For example, the incorporation of an incorrect nucleotide, or damage caused by mutagenic agents, can be corrected by cutting out a section of the daughter strand and recopying the parental strand.

Cell Division

Mitosis and Cytokinesis

In eukaryotes the processes of DNA replication and cell division occur at different times of the cell division cycle. During cell division, DNA condenses to form short, tightly coiled, rodlike chromosomes. Each chromosome then splits longitudinally, forming two identical chromatids. Each pair of chromatids is divided between the two daughter cells during mitosis, or division of the nucleus, a process in which the chromosomes are propelled by attachment to a bundle of microtubules called the mitotic spindle.

Mitosis can be divided into five phases. In prophase the mitotic spindle forms and the chromosomes condense. In prometaphase the nuclear envelope breaks down (in many but not all eukaryotes) and the chromosomes attach to the mitotic spindle. Both chromatids of each chromosome attach to the spindle at a specialized chromosomal region called the kinetochore. In metaphase the condensed chromosomes align in a plane across the equator of the mitotic spindle. Anaphase follows as the separated chromatids move abruptly toward opposite spindle poles. Finally, in telophase a new nuclear envelope forms around each set of unraveling chromatids.

An essential feature of mitosis is the attachment of the chromatids to opposite poles of the mitotic spindle. This ensures that each of the daughter cells will receive a complete set of chromosomes. The mitotic spindle is composed of microtubules, each of which is a tubular assembly of molecules of the protein tubulin. Some microtubules extend from one spindle pole to the other, while a second class extends from one spindle pole to a chromatid. Microtubules can grow or shrink by the addition or removal of tubulin molecules. The shortening of spindle microtubules at anaphase propels attached chromatids to the spindle poles, where they unravel to form new nuclei.

The two poles of the mitotic spindle are occupied by centrosomes, which organize the microtubule arrays. In animal cells each centrosome contains a pair of cylindrical centrioles, which are themselves composed of complex arrays of microtubules. Centrioles duplicate at a precise time in the cell division cycle, usually close to the start of DNA replication.

After mitosis comes cytokinesis, the division of the cytoplasm. This is another process in which animal and plant cells differ. In animal cells cytokinesis is achieved through the constriction of the cell by a ring of contractile microfilaments consisting of actin and myosin, the proteins involved in muscle contraction and other forms of cell movement. In plant cells the cytoplasm is divided by the formation of a new cell wall, called the cell plate, between the two daughter cells. The cell plate arises from small Golgi-derived vesicles that coalesce in a plane across the equator of the late telophase spindle to form a disk-shaped structure. In this process, each vesicle contributes its membrane to the forming cell membranes and its matrix contents to the forming cell wall. A second set of vesicles extends the edge of the cell plate until it reaches and fuses with the sides of the parent cell, thereby completely separating the two new daughter cells. At this point, cellulose synthesis commences, and the cell plate becomes a primary cell wall.

Meiosis

A specialized division of chromosomes called meiosis occurs during the formation of the reproductive cells, or gametes, of sexually reproducing organisms. Gametes such as ova, sperm, and pollen begin as germ cells, which, like other types of cells, have two copies of each gene in their nuclei. The chromosomes composed of these matching genes are called homologs. During DNA replication, each chromosome duplicates into two attached chromatids. The homologous chromosomes are then separated to opposite poles of the meiotic spindle by microtubules similar to those of the mitotic spindle. At this stage in the meiosis of germ cells, there is a crucial difference from the mitosis of other cells. In meiosis the two chromatids making up each chromosome remain together, so that whole chromosomes are separated from their homologous partners. Cell division then occurs, followed by a second division that resembles mitosis more closely in that it separates the two chromatids of each remaining chromosome. In this way, when meiosis is complete, each mature gamete receives only one copy of each gene instead of the two copies present in other cells.

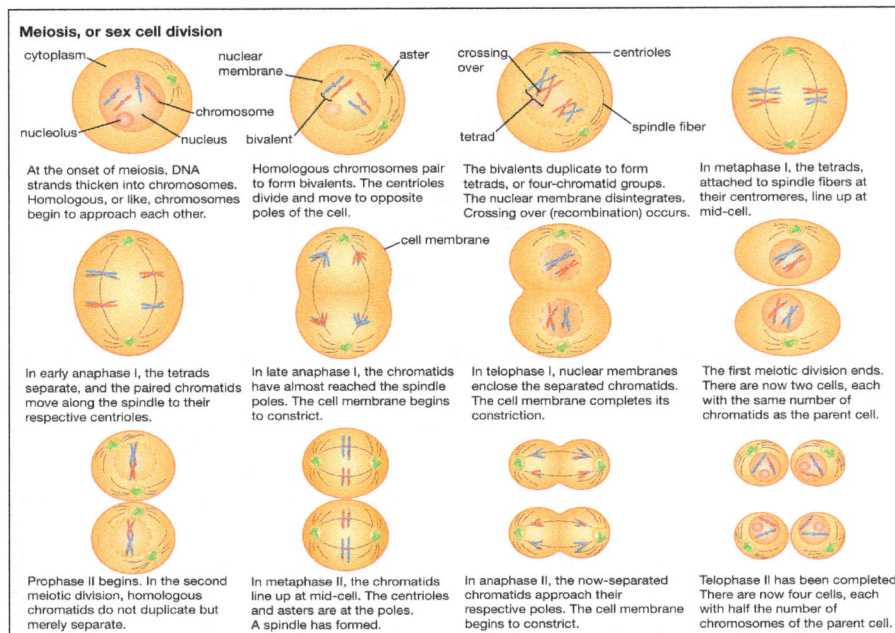

Meiosis, or sex cell division

At the onset of meiosis, DNA strands thicken into chromosomes. Homologous, or like, chromosomes begin to approach each other.

Homologous chromosomes pair to form bivalents. The centrioles divide and move to opposite poles of the cell.

The bivalents duplicate to form tetrads, or four-chromatid groups. The nuclear membrane disintegrates. Crossing over (recombination) occurs.

In metaphase I, the tetrads, attached to spindle fibers at their centromeres, line up at mid-cell.

In early anaphase I, the tetrads separate, and the paired chromatids move along the spindle to their respective centrioles.

In late anaphase I, the chromatids have almost reached the spindle poles. The cell membrane begins to constrict.

In telophase I, nuclear membranes enclose the separated chromatids. The cell membrane completes its constriction.

The first meiotic division ends. There are now two cells, each with the same number of chromatids as the parent cell.

Prophase II begins. In the second meiotic division, homologous chromatids do not duplicate but merely separate.

In metaphase II, the chromatids line up at mid-cell. The centrioles and asters are at the poles. A spindle has formed.

In anaphase II, the now-separated chromatids approach their respective poles. The cell membrane begins to constrict.

Telophase II has been completed. There are now four cells, each with half the number of chromosomes of the parent cell.

The formation of gametes (sex cells) occurs during the process of meiosis.

The Cell Division Cycle

In prokaryotes, DNA synthesis can take place uninterrupted between cell divisions, and new cycles of DNA synthesis can begin before previous cycles have finished. In contrast, eukaryotes duplicate their DNA exactly once during a discrete period between cell divisions. This period is called the S (for synthetic) phase. It is preceded by a period called G_1 (meaning "first gap") and followed by a period called G_2, during which nuclear DNA synthesis does not occur.

The four periods G_1, S, G_2, and M (for mitosis) make up the cell division cycle. The cell cycle characteristically lasts between 10 and 20 hours in rapidly proliferating adult cells, but it can be arrested for weeks or months in quiescent cells or for a lifetime in neurons of the brain. Prolonged arrest of this type usually occurs during the G_1 phase and is sometimes referred to as G0. In contrast, some embryonic cells, such as those of fruit flies (vinegar flies), can complete entire cycles and divide in only 11 minutes. In these exceptional cases, G_1 and G_2 are undetectable, and mitosis alternates with DNA synthesis. In addition, the duration of the S phase varies dramatically. The fruit fly embryo takes only four minutes to replicate its DNA, compared with several hours in adult cells of the same species.

Controlled Proliferation

Several studies have identified the transition from the G_1 to the S phase as a crucial control point of the cell cycle. Stimuli are known to cause resting cells to proliferate by inducing them to leave G_1 and begin DNA synthesis. These stimuli, called growth factors, are naturally occurring proteins specific to certain groups of cells in the body. They include nerve growth factor, epidermal growth factor, and platelet-derived growth factor. Such factors may have important roles in the healing of wounds as well as in the maintenance and growth of normal tissues. Many growth factors are known to act on the external membrane of the cell, by interacting with specialized protein receptor molecules. These respond by triggering further cellular changes, including an increase in calcium levels that makes the cell interior more alkaline and the addition of phosphate groups to the amino acid tyrosine in proteins. The complex response of cells to growth factors is of fundamental importance to the control of cell proliferation.

Failure of Proliferation Control

Cancer can arise when the controlling factors over cell growth fail and allow a cell and its descendants to keep dividing at the expense of the organism. Studies of viruses that transform cultured cells and thus lead to the loss of control of cell growth have provided insight into the mechanisms that drive the formation of tumours. Transformed cells may differ from their normal progenitors by continuing to proliferate at very high densities, in the absence of growth factors, or in the absence of a solid substrate for support.

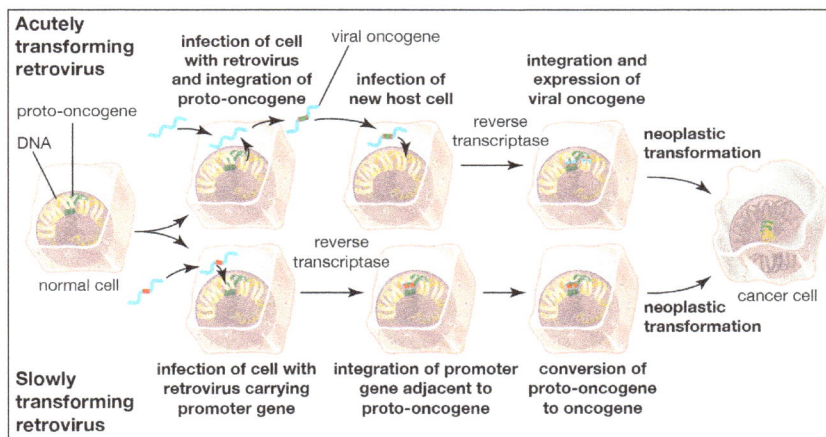

Cancer-causing retroviruses.

In figure, retroviral insertion can convert a proto-oncogene, integral to the control of cell division, into an oncogene, the agent responsible for transforming a healthy cell into a cancer cell. An acutely transforming retrovirus (shown at top), which produces tumours within weeks of infection, incorporates genetic material from a host cell into its own genome upon infection, forming a viral oncogene. When the viral oncogene infects another cell, an enzyme called reverse transcriptase copies the single-stranded genetic material into double-stranded DNA, which is then integrated into the cellular genome. A slowly transforming retrovirus (shown at bottom), which requires months to elicit tumour growth, does not disrupt cellular function through the insertion of a viral oncogene. Rather, it carries a promoter gene that is integrated into the cellular genome of the host cell next to or within a proto-oncogene, allowing conversion of the proto-oncogene to an oncogene.

Major advances in the understanding of growth control have come from studies of the viral genes that cause transformation. These viral oncogenes have led to the identification of related cellular genes called protooncogenes. Protooncogenes can be altered by mutation or epigenetic modification, which converts them into oncogenes and leads to cell transformation. Specific oncogenes are activated in particular human cancers. For example, an oncogene called RAS is associated with many epithelial cancers, while another, called MYC, is associated with leukemias.

An interesting feature of oncogenes is that they may act at different levels corresponding to the multiple steps seen in the development of cancer. Some oncogenes immortalize cells so that they divide indefinitely, whereas normal cells die after a limited number of generations. Other oncogenes transform cells so that they grow in the absence of growth factors. A combination of these two functions leads to loss of proliferation control, whereas each of these functions on its own cannot. The mode of action of oncogenes also provides important clues to the nature of growth control and cancer. For example, some oncogenes are known to encode receptors for growth factors that may cause continuous proliferation in the absence of appropriate growth factors.

Loss of growth control has the added consequence that cells no longer repair their DNA effectively, and thus aberrant mitoses occur. As a result, additional mutations arise that subvert a cell's normal constraints to remain in its tissue of origin. Epithelial tumour cells, for example, acquire the ability to cross the basal lamina and enter the bloodstream or lymphatic system, where they migrate to other parts of the body, a process called metastasis. When cells metastasize to distant tissues, the tumour is described as malignant, whereas prior to metastasis a tumour is described as benign.

Cell Differentiation

Adult organisms are composed of a number of distinct cell types. Cells are organized into tissues, each of which typically contains a small number of cell types and is devoted to a specific physiological function. For example, the epithelial tissue lining the small intestine contains columnar absorptive cells, mucus-secreting goblet cells, hormone-secreting endocrine cells, and enzyme-secreting Paneth cells. In addition, there exist undifferentiated dividing cells that lie in the crypts between the intestinal villi and serve to replace the other cell types when they become damaged or worn out. Another example of a differentiated tissue is the skeletal tissue of a long bone, which contains osteoblasts (large cells that synthesize bone) in the outer sheath and osteocytes (mature bone cells) and osteoclasts (multinucleate cells involved in bone remodeling) within the matrix.

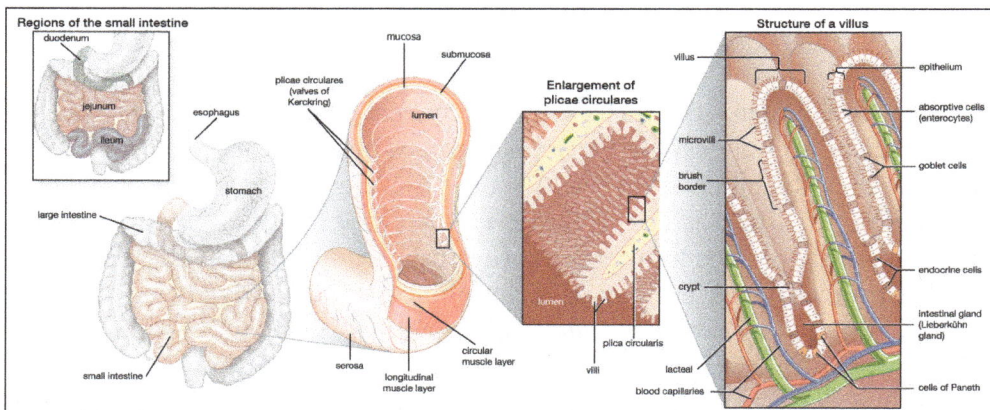

The small intestine contains many distinct types of cells, each of which serves a specific function.

In general, the simpler the overall organization of the animal, the fewer the number of distinct cell types that they possess. Mammals contain more than 200 different cell types, whereas simple invertebrate animals may have only a few different types. Plants are also made up of differentiated cells, but they are quite different from the cells of animals. For example, a leaf in a higher plant is covered with a cuticle layer of epidermal cells. Among these are pores composed of two specialized cells, which regulate gaseous exchange across the epidermis. Within the leaf is the mesophyll, a spongy tissue responsible for photosynthetic activity. There are also veins composed of xylem elements, which transport water up from the soil, and phloem elements, which transport products of photosynthesis to the storage organs.

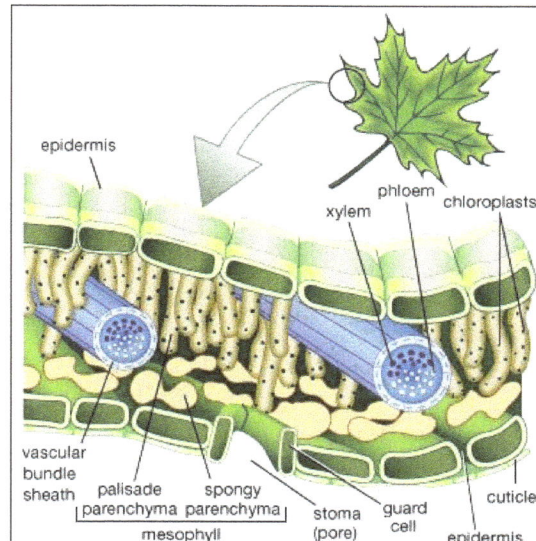

Structures of a leafThe epidermis is often covered with a waxy protective cuticle that helps prevent water loss from inside the leaf. Oxygen, carbon dioxide, and water enter and exit the leaf through pores (stomata) scattered mostly along the lower epidermis. The stomata are opened and closed by the contraction and expansion of surrounding guard cells. The vascular, or conducting, tissues are known as xylem and phloem; water and minerals travel up to the leaves from the roots through the xylem, and sugars made by photosynthesis are transported to other parts of the plant through the phloem. Photosynthesis occurs within the chloroplast-containing mesophyll layer.

The various cell types have traditionally been recognized and classified according to their appearance in the light microscope following the process of fixing, processing, sectioning, and staining tissues that is known as histology. Classical histology has been augmented by a variety of more discriminating techniques. Electron microscopy allows for higher magnifications. Histochemistry involves the use of coloured precipitating substrates to stain particular enzymes in situ. Immunohistochemistry uses specific antibodies to identify particular substances, usually proteins or carbohydrates, within cells. In situ hybridization involves the use of nucleic acid probes to visualize the location of specific messenger RNAs (mRNA). These modern methods have allowed the identification of more cell types than could be visualized by classical histology, particularly in the brain, the immune system, and among the hormone-secreting cells of the endocrine system.

The Differentiated State

The biochemical basis of cell differentiation is the synthesis by the cell of a particular set of proteins, carbohydrates, and lipids. This synthesis is catalyzed by proteins called enzymes. Each enzyme in turn is synthesized in accordance with a particular gene, or sequence of nucleotides in the DNA of the cell nucleus. A particular state of differentiation, then, corresponds to the set of genes that is expressed and the level to which it is expressed.

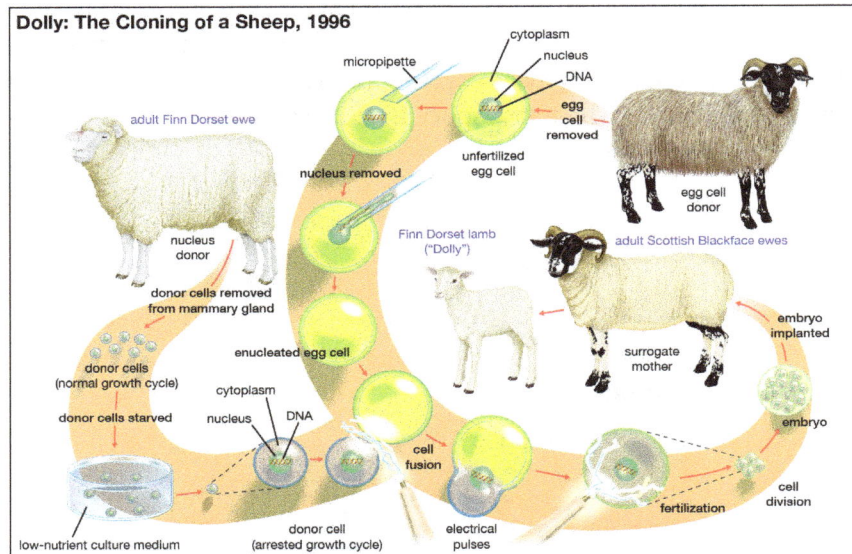

Dolly: The Cloning of a Sheep, 1996

Dolly the sheep was successfully cloned in 1996 by fusing the nucleus from a mammary-gland cell of a Finn Dorset ewe into an enucleated egg cell taken from a Scottish Blackface ewe. Carried to term in the womb of another Scottish Blackface ewe, Dolly was a genetic copy of the Finn Dorset ewe.

It is believed that all of an organism's genes are present in each cell nucleus, no matter what the cell type, and that differences between tissues are not due to the presence or absence of certain genes but are due to the expression of some and the repression of others. In animals the best evidence for retention of the entire set of genes comes from whole animal cloning experiments in which the nucleus of a differentiated cell is substituted for the nucleus of a fertilized egg. In many species this can result in the development of a normal embryo that contains the full range of body parts and cell types. Likewise, in plants it is often possible to grow complete embryos from individual cells in tissue culture. Such experiments show that any nucleus has the genetic information required for the growth of a developing organism, and they strongly suggest that, for most tissues, cell differentiation arises from the regulation of genetic activity rather than the removal or destruction of unwanted genes. The only known exception to this rule comes from the immune system, where segments of DNA in developing white blood cells are slightly rearranged, producing a wide variety of antibody and receptor molecules.

At the molecular level there are many ways in which the expression of a gene can be differentially regulated in different cell types. There may be differences in the copying, or transcription, of the gene into RNA; in the processing of the initial RNA transcript into mRNA; in the control of mRNA movement to the cytoplasm; in the translation of mRNA to protein; or in the stability of mRNA. However, the control of transcription has the most influence over gene expression and has received the most detailed analysis.

The DNA in the cell nucleus exists in the form of chromatin, which is made up of DNA bound to histones (simple alkaline proteins) and other nonhistone proteins. Most of the DNA is complexed into repeating structures called nucleosomes, each of which contains eight molecules of histone.

Active genes are found in parts of the DNA where the chromatin has an "open" configuration, in which regulatory proteins are able to gain access to the DNA. The degree to which the chromatin opens depends on chemical modifications of the outer parts of the histone molecules and on the presence or absence of particular nonhistone proteins. Transcriptional control is exerted with the help of regulatory sequences that are found associated with a gene, such as the promoter sequence, a region near the start of the gene, and enhancer sequences, regions that lie elsewhere within the DNA that augment the activity of enzymes involved in the process of transcription. Whether or not transcription occurs depends on the binding of transcription factors to these regulatory sequences. Transcription factors are proteins that usually possess a DNA-binding region, which recognizes the specific regulatory sequence in the DNA, and an effector region, which activates or inhibits transcription. Transcription factors often work by recruiting enzymes that add modifications (e.g., acetyl groups or methyl groups) to or remove modifications from the outer parts of the histone molecules. This controls the folding of the chromatin and the accessibility of the DNA to RNA polymerase and other transcription factors.

In general, it requires several transcription factors working in combination to activate a gene. For example, the chicken delta 1 crystallin gene, normally expressed only in the lens of the eye, has a promoter that contains binding sites for two activating transcription factors and an enhancer that contains binding sites for two other activating transcription factors. There is also an additional enhancer site that can bind either an activator (deltaEF3) or a repressor (deltaEF1). Successful transcription requires that all these sites are occupied by the correct transcription factors.

Fully differentiated cells are qualitatively different from one another. States of terminal differentiation are stable and persistent, both in the lifetime of the cell and in successive cell generations (in the case of differentiated types that are capable of continued cell division). The inherent stability of the differentiated state is maintained by various processes, including feedback activation of genes by their own products and repression of inactive genes. Chromatin structure may be important in maintaining states of differentiation, although it is still unclear whether this can be maintained during DNA replication, which involves temporary removal of chromosomal proteins and unwinding of the DNA double helix.

A type of differentiation control that is maintained during DNA replication is the methylation of DNA, which tends to recruit histone deacetylases and hence close up the structure of the chromatin. DNA methylation occurs when a methyl group is attached to the exterior, or sugar-phosphate side, of a cytosine (C) residue. Cytosine methylation occurs only on a C nucleotide when it is connected to a G (guanine) nucleotide on the same strand of DNA. These nucleotide pairings are called CG dinucleotides. One class of DNA methylase enzyme can introduce new methylations when required, whereas another class, called maintenance methylases, methylates CG dinucleotides in the DNA double helix only when the CG of the complementary strand is already methylated. Each time the methylated DNA is replicated, the old strand has the methyl groups and the new strand does not. The maintenance methylase will then add methyl groups to all the CGs opposite the existing methyl groups to restore a fully methylated double helix. This mechanism guarantees stability of the DNA methylation pattern, and hence the differentiated state, during the processes of DNA replication and cell division.

The Process of Differentiation

Differentiation from visibly undifferentiated precursor cells occurs during embryonic development, during metamorphosis of larval forms, and following the separation of parts in asexual reproduction. It also takes place in adult organisms during the renewal of tissues and the regeneration of missing parts. Thus, cell differentiation is an essential and ongoing process at all stages of life.

The visible differentiation of cells is only the last of a progressive sequence of states. In each state, the cell becomes increasingly committed toward one type of cell into which it can develop. States of commitment are sometimes described as "specification" to represent a reversible type of commitment and as "determination" to represent an irreversible commitment. Although states of specification and determination both represent differential gene activity, the properties of embryonic cells are not necessarily the same as those of fully differentiated cells. In particular, cells in specification states are usually not stable over prolonged periods of time.

Two mechanisms bring about altered commitments in the different regions of the early embryo: cytoplasmic localization and induction. Cytoplasmic localization is evident in the earliest stages of development of the embryo. During this time, the embryo divides without growth, undergoing cleavage divisions that produce separate cells called blastomeres. Each blastomere inherits a certain region of the original egg cytoplasm, which may contain one or more regulatory substances called cytoplasmic determinants. When the embryo has become a solid mass of blastomeres (called a morula), it generally consists of two or more differently committed cell populations—a result of the blastomeres having incorporated different cytoplasmic determinants. Cytoplasmic determinants may consist of mRNA or protein in a particular state of activation. An example of the influence of a cytoplasmic determinant is a receptor called Toll, located in the membranes of Drosophila (fruit fly) eggs. Activation of Toll ensures that the blastomeres will develop into ventral (underside) structures, while blastomeres containing inactive Toll will produce cells that will develop into dorsal (back) structures.

In induction, the second mechanism of commitment, a substance secreted by one group of cells alters the development of another group. In early development, induction is usually instructive; that is, the tissue assumes a different state of commitment in the presence of the signal than it would in the absence of the signal. Inductive signals often take the form of concentration gradients of substances that evoke a number of different responses at different concentrations. This leads to the formation of a sequence of groups of cells, each in a different state of specification. For example, in Xenopus (clawed frog) the early embryo contains a signaling centre called the organizer that secretes inhibitors of bone morphogenetic proteins (BMPs), leading to a ventral-to-dorsal (belly-to-back) gradient of BMP activity. The activity of BMP in the ventral region of the embryo suppresses the expression of transcription factors involved in the formation of the central nervous system and segmented muscles. Suppression ensures that these structures are formed only on the dorsal side, where there is decreased activity of BMP.

The final stage of differentiation often involves the formation of several types of differentiated cells from one precursor or stem cell population. Terminal differentiation occurs not only in embryonic development but also in many tissues in postnatal life. Control of this process depends on a system of lateral inhibition in which cells that are differentiating along a particular pathway send out signals that repress similar differentiation by their neighbours. For example, in the developing central nervous system of vertebrates, neurons arise from a simple tube of neuroepithelium, the cells of which possess a surface receptor called Notch. These cells also possess another cell surface molecule called Delta that can bind to and activate Notch on adjacent cells. Activation of Notch initiates a cascade of intracellular events that results in suppression of Delta production and suppression of neuronal differentiation. This means that the neuroepithelium generates only a few cells with high expression of Delta surrounded by a larger number of cells with low expression of Delta. High Delta production and low Notch activation makes the cells develop into neurons. Low Delta production and high Notch activation makes the cells remain as precursor cells or become glial (supporting) cells. A similar mechanism is known to produce the endocrine cells of the pancreas and the goblet cells of the intestinal epithelium. Such lateral inhibition systems work because cells in a population are never quite identical to begin with. There are always small differences, such as in the number of Delta molecules displayed on the cell surface. The mechanism of lateral inhibition amplifies these small differences, using them to bring about differential gene expression that leads to stable and persistent states of cell differentiation.

Errors in Differentiation

Three classes of abnormal cell differentiation are dysplasia, metaplasia, and anaplasia. Dysplasia indicates an abnormal arrangement of cells, usually arising from a disturbance in their normal growth behaviour. Some dysplasias are precursor lesions to cancer, whereas others are harmless and regress spontaneously. For example, dysplasia of the uterine cervix, called cervical intraepithelial neoplasia (CIN), may progress to cervical cancer. It can be detected by cervical smear cytology tests (Pap smears).

Metaplasia is the conversion of one cell type into another. In fact, it is not usually the differentiated cells themselves that change but rather the stem cell population from which they are derived. Metaplasia commonly occurs where chronic tissue damage is followed by extensive regeneration. For example, squamous metaplasia of the bronchi occurs when the ciliated respiratory epithelial cells of people who smoke develop into squamous, or flattened, cells. In intestinal metaplasia of the stomach, patches resembling intestinal tissue arise in the gastric mucosa, often in association with gastric ulcers. Both of these types of metaplasia may progress to cancer.

Anaplasia is a loss of visible differentiation that can occur in advanced cancer. In general, early cancers resemble their tissue of origin and are described and classified by their pattern of differentiation. However, as they develop, they produce variants of more abnormal appearance and increased malignancy. Finally, a highly anaplastic growth can occur, in which the cancerous cells bear no visible relation to the parent tissue.

The Evolution of Cells

The Development of Genetic Information

Life on Earth could not exist until a collection of catalysts appeared that could promote the synthesis of more catalysts of the same kind. Early stages in the evolutionary pathway of cells presumably centred on RNA molecules, which not only present specific catalytic surfaces but also contain the potential for their own duplication through the formation of a complementary RNA molecule. It is assumed that a small RNA molecule eventually appeared that was able to catalyze its own duplication.

Imperfections in primitive RNA replication likely gave rise to many variant autocatalytic RNA molecules. Molecules of RNA that acquired variations that increased the speed or the fidelity of self-replication would have outmultiplied other, less-competent RNA molecules. In addition, other small RNA molecules that existed in symbiosis with autocatalytic RNA molecules underwent natural selection for their ability to catalyze useful secondary reactions such as the production of better precursor molecules. In this way, sophisticated families of RNA catalysts could have evolved together, since cooperation between different molecules produced a system that was much more effective at self-replication than a collection of individual RNA catalysts.

Another major step in the evolution of the cell would have been the development, in one family of self-replicating RNA, of a primitive mechanism of protein synthesis. Protein molecules cannot provide the information for the synthesis of other protein molecules like themselves. This information must ultimately be derived from a nucleic acid sequence. Protein synthesis is much more complex than RNA synthesis, and it could not have arisen before a group of powerful RNA catalysts evolved. Each of these catalysts presumably has its counterpart among the RNA molecules that function in the current cell: (1) there was an information RNA molecule, much like messenger RNA (mRNA), whose nucleotide sequence was read to create an amino acid sequence; (2) there was a group of adaptor RNA molecules, much like transfer RNA (tRNA), that could bind to both mRNA and a specific activated amino acid; and (3) finally, there was an RNA catalyst, much like ribosomal RNA (rRNA), that facilitated the joining together of the amino acids aligned on the mRNA by the adaptor RNA.

At some point in the evolution of biological catalysts, the first cell was formed. This would have required the partitioning of the primitive biological catalysts into individual units, each surrounded by a membrane. Membrane formation might have occurred quite simply, since many amphiphilic molecules—half hydrophobic (water-repelling) and half hydrophilic (water-loving)—aggregate to form bilayer sheets in which the hydrophobic portions of the molecules line up in rows to form the interior of the sheet and leave the hydrophilic portions to face the water. Such bilayer sheets can spontaneously close up to form the walls of small, spherical vesicles, as can the phospholipid bilayer membranes of present-day cells.

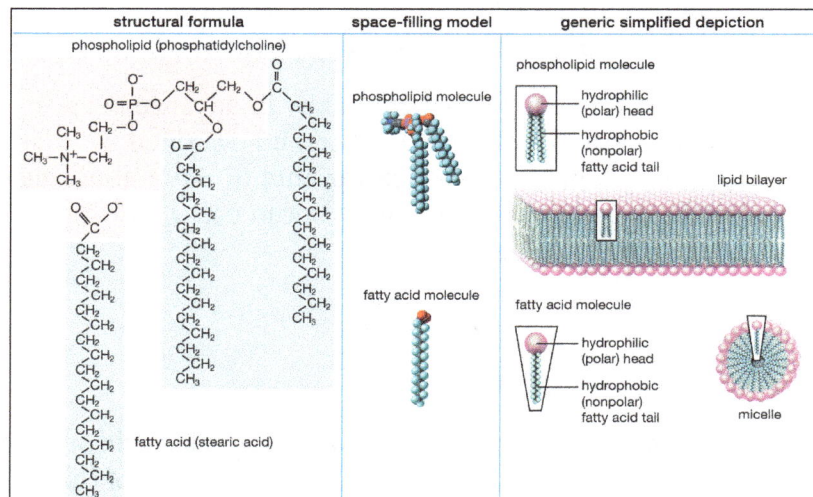

In figure, structure and properties of two representative lipidsBoth stearic acid (a fatty acid) and phosphatidylcholine (a phospholipid) are composed of chemical groups that form polar "heads" and nonpolar "tails." The polar heads are hydrophilic, or soluble in water, whereas the nonpolar tails are hydrophobic, or insoluble in water. Lipid molecules of this composition spontaneously form aggregate structures such as micelles and lipid bilayers, with their hydrophilic ends oriented toward the watery medium and their hydrophobic ends shielded from the water.

As soon as the biological catalysts became compartmentalized into small individual units, or cells, the units would have begun to compete with one another for the same resources. The active competition that ensued must have greatly accelerated evolutionary change, serving as a powerful force for the development of more efficient cells. In this way, cells eventually arose that contained new catalysts, enabling them to use simpler, more abundant precursor molecules for their growth. Because these cells were no longer dependent on preformed ingredients for their survival, they were able to spread far beyond the limited environments where the first primitive cells arose.

It is often assumed that the first cells appeared only after the development of a primitive form of protein synthesis. However, it is by no means certain that cells cannot exist without proteins, and it has been suggested that the first cells contained only RNA catalysts. In either case, protein molecules, with their chemically varied side chains, are more powerful catalysts than RNA molecules; therefore, as time passed, cells arose in which RNA served primarily as genetic material, being directly replicated in each generation and inherited by all progeny cells in order to specify proteins.

As cells became more complex, a need would have arisen for a stabler form of genetic information storage than that provided by RNA. DNA, related to RNA yet chemically stabler, probably appeared rather late in the evolutionary history of cells. Over a period of time, the genetic information in RNA sequences was transferred to DNA sequences, and the ability of RNA molecules to replicate directly was lost. It was only at this point that the central process of biology—the synthesis, one after the other, of DNA, RNA, and protein—appeared.

The Development of Metabolism

The first cells presumably resembled prokaryotic cells in lacking nuclei and functional internal compartments, or organelles. These early cells were also anaerobic (not requiring oxygen), deriving their energy from the fermentation of organic molecules that had previously accumulated on the Earth over long periods of time. Eventually, more sophisticated cells evolved that could carry out primitive forms of photosynthesis, in which light energy was harnessed by membrane-bound proteins to form organic molecules with energy-rich chemical bonds. A major turning point in the evolution of life was the development of photosynthesizing prokaryotes requiring only water as an electron donor and capable of producing molecular oxygen. The descendants of these prokaryotes, the blue-green algae (cyanobacteria), still exist as viable life-forms. Their ancestors prospered to such an extent that the atmosphere became rich in the oxygen they produced. The free availability of this oxygen in turn enabled other prokaryotes to evolve aerobic forms of metabolism that were much more efficient in the use of organic molecules as a source of food.

The switch to predominantly aerobic metabolism is thought to have occurred in bacteria approximately 2 billion years ago, about 1.5 billion years after the first cells had formed. Aerobic eukaryotic cells (cells containing nuclei and all the other organelles) probably appeared about 1.5 billion years ago, their lineage having branched off much earlier from that of the prokaryotes. Eukaryotic cells almost certainly became aerobic by engulfing aerobic prokaryotes, with which they lived in a symbiotic relationship. The mitochondria found in both animals and plants are the descendants of such prokaryotes. Later, in branches of the eukaryotic lineage leading to plants and algae, a blue-green algaelike organism was engulfed to perform photosynthesis. It is likely that over a long period of time these organisms became the chloroplasts.

The eukaryotic cell thus apparently arose as an amalgam of different cells, in the process becoming an efficient aerobic cell whose plasma membrane was freed from energy metabolism—one of the major functions of the cell membrane of prokaryotes. The eukaryotic cell membrane was therefore able to become specialized for cell-to-cell communication and cell signaling. It may be partly for this reason that eukaryotic cells were eventually more successful at forming complex multicellular organisms than their simpler prokaryotic relatives.

Gene

Gene is the unit of hereditary information that occupies a fixed position (locus) on a chromosome. Genes achieve their effects by directing the synthesis of proteins.

In eukaryotes (such as animals, plants, and fungi), genes are contained within the cell nucleus. The mitochondria (in animals) and the chloroplasts (in plants) also contain small subsets of genes distinct from the genes found in the nucleus. In prokaryotes (organisms lacking a distinct nucleus, such as bacteria), genes are contained in a single chromosome that is free-floating in the cell cytoplasm. Many bacteria also contain plasmids—extrachromosomal genetic elements with a small number of genes.

The number of genes in an organism's genome (the entire set of chromosomes) varies significantly between species. For example, whereas the human genome contains an estimated 20,000 to 25,000 genes, the genome of the bacterium Escherichia coli O157:H7 houses precisely 5,416 genes. Arabidopsis thaliana—the first plant for which a complete genomic sequence was recovered—has roughly 25,500 genes; its genome is one of the smallest known to plants. Among extant independently replicating organisms, the bacterium Mycoplasma genitalium has the fewest number of genes, just 517.

Chemical Structure of Genes

Genes are composed of deoxyribonucleic acid (DNA), except in some viruses, which have genes consisting of a closely related compound called ribonucleic acid (RNA). A DNA molecule is composed of two chains of nucleotides that wind about each other to resemble a twisted ladder. The sides of the ladder are made up of sugars and phosphates, and the rungs are formed by bonded pairs of nitrogenous bases. These bases are adenine (A), guanine (G), cytosine (C), and thymine (T). An A on one chain bonds to a T on the other (thus forming an A–T ladder rung); similarly, a C on one chain bonds to a G on the other. If the bonds between the bases are broken, the two chains unwind, and free nucleotides within the cell attach themselves to the exposed bases of the now-separated chains. The free nucleotides line up along each chain according to the base-pairing rule—A bonds to T, C bonds to G. This process results in the creation of two identical DNA molecules from one original and is the method by which hereditary information is passed from one generation of cells to the next.

Gene Transcription and Translation

The sequence of bases along a strand of DNA determines the genetic code. When the product of a particular gene is needed, the portion of the DNA molecule that contains that gene will split.

Through the process of transcription, a strand of RNA with bases complementary to those of the gene is created from the free nucleotides in the cell. (RNA has the base uracil [U] instead of thymine, so A and U form base pairs during RNA synthesis.) This single chain of RNA, called messenger RNA (mRNA), then passes to the organelles called ribosomes, where the process of translation, or protein synthesis, takes place. During translation, a second type of RNA, transfer RNA (tRNA), matches up the nucleotides on mRNA with specific amino acids. Each set of three nucleotides codes for one amino acid. The series of amino acids built according to the sequence of nucleotides forms a polypeptide chain; all proteins are made from one or more linked polypeptide chains.

Experiments conducted in the 1940s indicated one gene being responsible for the assembly of one enzyme, or one polypeptide chain. This is known as the one gene–one enzyme hypothesis. However, since this discovery, it has been realized that not all genes encode an enzyme and that some enzymes are made up of several short polypeptides encoded by two or more genes.

Gene Regulation

Experiments have shown that many of the genes within the cells of organisms are inactive much or even all of the time. Thus, at any time, in both eukaryotes and prokaryotes, it seems that a gene can be switched on or off. The regulation of genes between eukaryotes and prokaryotes differs in important ways.

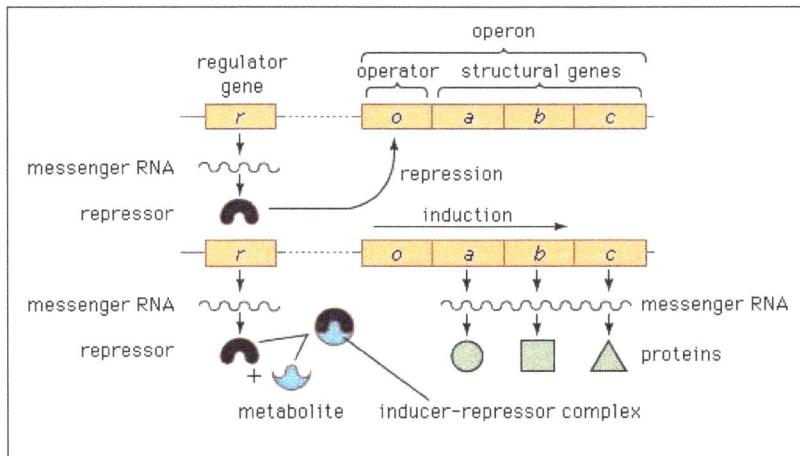

Model of the operon and its regulator gene.

The process by which genes are activated and deactivated in bacteria is well characterized. Bacteria have three types of genes: structural, operator, and regulator. Structural genes code for the synthesis of specific polypeptides. Operator genes contain the code necessary to begin the process of transcribing the DNA message of one or more structural genes into mRNA. Thus, structural genes are linked to an operator gene in a functional unit called an operon. Ultimately, the activity of the operon is controlled by a regulator gene, which produces a small protein molecule called a repressor. The repressor binds to the operator gene and prevents it from initiating the synthesis of the protein called for by the operon. The presence or absence of certain repressor molecules determines whether the operon is off or on. As mentioned, this model applies to bacteria.

The genes of eukaryotes, which do not have operons, are regulated independently. The series of events associated with gene expression in higher organisms involves multiple levels of regulation and is often influenced by the presence or absence of molecules called transcription factors. These factors influence the fundamental level of gene control, which is the rate of transcription, and may function as activators or enhancers. Specific transcription factors regulate the production of RNA from genes at certain times and in certain types of cells. Transcription factors often bind to the promoter, or regulatory region, found in the genes of higher organisms. Following transcription, introns (noncoding nucleotide sequences) are excised from the primary transcript through processes known as editing and splicing. The result of these processes is a functional strand of mRNA. For most genes this is a routine step in the production of mRNA, but in some genes there are multiple ways to splice the primary transcript, resulting in different mRNAs, which in turn result in different proteins. Some genes also are controlled at the translational and posttranslational levels.

Gene Mutations

Mutations occur when the number or order of bases in a gene is disrupted. Nucleotides can be deleted, doubled, rearranged, or replaced, each alteration having a particular effect. Mutation generally has little or no effect, but, when it does alter an organism, the change may be lethal or cause disease. A beneficial mutation will rise in frequency within a population until it becomes the norm.

Allele

An allele is a variant form of a given gene. Sometimes, different alleles can result in different observable phenotypic traits, such as different pigmentation. A notable example of this trait of color variation is Gregor Mendel's discovery that the white and purple flower colors in pea plants were the result of "pure line" traits which could be used as a control for future experiments. However, most genetic variations result in little or no observable variation.

Most multicellular organisms have two sets of chromosomes; that is, they are diploid. In this case, the chromosomes can be paired: each pair is made up of two chromosomes of the same type, known as homologous chromosomes. If both alleles at a gene (or locus) on the homologous chromosomes are the same, they and the organism are homozygous with respect to that gene (or locus). If the alleles are different, they and the organism are heterozygous with respect to that gene.

Alleles that Lead to Dominant or Recessive Phenotypes

In many cases, genotypic interactions between the two alleles at a locus can be described as dominant or recessive, according to which of the two homozygous phenotypes the heterozygote most resembles. Where the heterozygote is indistinguishable from one of the homozygotes, the allele expressed is the one that leads to the "dominant" phenotype, and the other allele is said to be "recessive". The degree and pattern of dominance varies among loci. This type of interaction was first formally described by Gregor Mendel. However, many traits defy this simple categorization and the phenotypes are modeled by co-dominance and polygenic inheritance.

The term "wild type" allele is sometimes used to describe an allele that is thought to contribute to the typical phenotypic character as seen in "wild" populations of organisms, such as fruit flies (*Drosophila melanogaster*). Such a "wild type" allele was historically regarded as leading to a dominant (overpowering - always expressed), common, and normal phenotype, in contrast to "mutant" alleles that lead to recessive, rare, and frequently deleterious phenotypes. It was formerly thought that most individuals were homozygous for the "wild type" allele at most gene loci, and that any alternative "mutant" allele was found in homozygous form in a small minority of "affected" individuals, often as genetic diseases, and more frequently in heterozygous form in "carriers" for the mutant allele. It is now appreciated that most or all gene loci are highly polymorphic, with multiple alleles, whose frequencies vary from population to population, and that a great deal of genetic variation is hidden in the form of alleles that do not produce obvious phenotypic differences.

Multiple Alleles

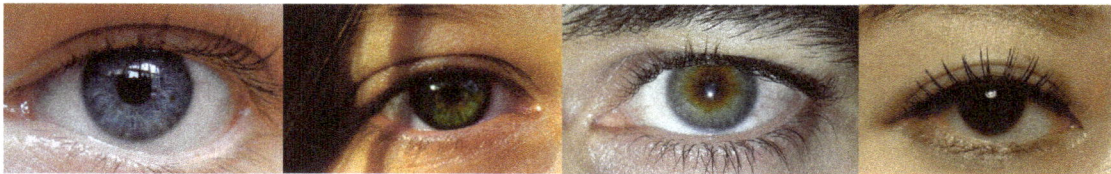

Eye color is an inherited trait influenced by more than one gene, including *OCA2* and *HERC2*. The interaction of multiple genes—and the variation in these genes ("alleles") between individuals—help to determine a person's eye color phenotype. Eye color is influenced by pigmentation of the iris and the frequency-dependence of the light scattering by the turbid medium within the stroma of the iris.

	Group A	Group B	Group AB	Group O
Red blood cell type	A	B	AB	O
Antibodies in plasma	Anti-B	Anti-A	None	Anti-A and Anti-B
Antigens in red blood cell	A antigen	B antigen	A and B antigens	None

In the ABO blood group system, a person with Type A blood displays A-antigens and may have a genotype I^AI^A or I^Ai. A person with Type B blood displays B-antigens and may have the genotype I^BI^B or I^Bi. A person with Type AB blood displays both A- and B-antigens and has the genotype I^AI^B and a person with Type O blood, displaying neither antigen, has the genotype ii.

A population or species of organisms typically includes multiple alleles at each locus among various individuals. Allelic variation at a locus is measurable as the number of alleles (polymorphism) present, or the proportion of heterozygotes in the population. A null allele is a gene variant that lacks the gene's normal function because it either is not expressed, or the expressed protein is inactive.

For example, at the gene locus for the ABO blood type carbohydrate antigens in humans, classical genetics recognizes three alleles, I^A, I^B, and i, which determine compatibility of blood transfusions. Any individual has one of six possible genotypes (I^AI^A, I^Ai, I^BI^B, I^Bi, I^AI^B, and ii) which produce one of four possible phenotypes: "Type A" (produced by $I^A I^A$ homozygous and $I^A i$ heterozygous genotypes), "Type B" (produced by $I^B I^B$ homozygous and $I^B i$ heterozygous genotypes), "Type AB" produced by $I^A I^B$ heterozygous genotype, and "Type O" produced by ii homozygous genotype. (It is now known that each of the A, B, and O alleles is actually a class of multiple alleles with different DNA sequences that produce proteins with identical properties: more than 70 alleles are known at the ABO locus. Hence an individual with "Type A" blood may be an AO heterozygote, an AA homozygote, or an AA heterozygote with two different "A" alleles.)

Genotype Frequencies

The frequency of alleles in a diploid population can be used to predict the frequencies of the corresponding genotypes. For a simple model, with two alleles;

$$p + q = 1$$

$$p^2 + 2pq + q^2 = 1$$

where p is the frequency of one allele and q is the frequency of the alternative allele, which necessarily sum to unity. Then, p^2 is the fraction of the population homozygous for the first allele, $2pq$ is the fraction of heterozygotes, and q^2 is the fraction homozygous for the alternative allele. If the first allele is dominant to the second then the fraction of the population that will show the dominant phenotype is $p^2 + 2pq$, and the fraction with the recessive phenotype is q^2.

With three alleles:

$$p + q + r = 1 \text{ and } p^2 + q^2 + r^2 + 2pq + 2pr + 2qr = 1.$$

In the case of multiple alleles at a diploid locus, the number of possible genotypes (G) with a number of alleles (a) is given by the expression:

$$G = \frac{a(a+1)}{2}.$$

Allelic Dominance in Genetic Disorders

A number of genetic disorders are caused when an individual inherits two recessive alleles for a single-gene trait. Recessive genetic disorders include albinism, cystic fibrosis, galactosemia, phenylketonuria (PKU), and Tay–Sachs disease. Other disorders are also due to recessive alleles, but because the gene locus is located on the X chromosome, so that males have only one copy (that is, they are hemizygous), they are more frequent in males than in females. Examples include red-green color blindness and fragile X syndrome.

Other disorders, such as Huntington's disease, occur when an individual inherits only one dominant allele.

Epialleles

While heritable traits are typically studied in terms of genetic alleles, epigenetic marks such as

DNA methylation can be inherited at specific genomic regions in certain species, a process termed transgenerational epigenetic inheritance. The term *epiallele* is used to distinguish these heritable marks from traditional alleles, which are defined by nucleotide sequence. A specific class of epiallele, the metastable epialleles, has been discovered in mice and in humans which is characterized by stochastic (probabilistic) establishment of epigenetic state that can be mitotically inherited.

Proteins

Proteins are large biomolecules, or macromolecules, consisting of one or more long chains of amino acid residues. Proteins perform a vast array of functions within organisms, including catalysing metabolic reactions, DNA replication, responding to stimuli, providing structure to cells and organisms, and transporting molecules from one location to another. Proteins differ from one another primarily in their sequence of amino acids, which is dictated by the nucleotide sequence of their genes, and which usually results in protein folding into a specific three-dimensional structure that determines its activity.

A linear chain of amino acid residues is called a polypeptide. A protein contains at least one long polypeptide. Short polypeptides, containing less than 20–30 residues, are rarely considered to be proteins and are commonly called peptides, or sometimes oligopeptides. The individual amino acid residues are bonded together by peptide bonds and adjacent amino acid residues. The sequence of amino acid residues in a protein is defined by the sequence of a gene, which is encoded in the genetic code. In general, the genetic code specifies 20 standard amino acids; however, in certain organisms the genetic code can include selenocysteine and—in certain archaea—pyrrolysine. Shortly after or even during synthesis, the residues in a protein are often chemically modified by post-translational modification, which alters the physical and chemical properties, folding, stability, activity, and ultimately, the function of the proteins. Sometimes proteins have non-peptide groups attached, which can be called prosthetic groups or cofactors. Proteins can also work together to achieve a particular function, and they often associate to form stable protein complexes.

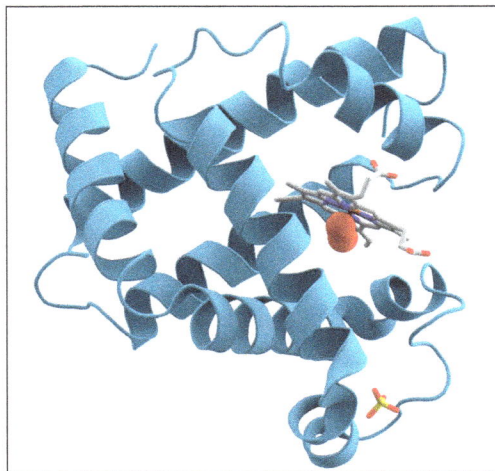

A representation of the 3D structure of the protein myoglobin showing turquoise α-helices. This protein was the first to have its structure solved by X-ray crystallography. Towards the right-center among the coils, a prosthetic group called a heme group (shown in gray) with a bound oxygen molecule (red).

Once formed, proteins only exist for a certain period and are then degraded and recycled by the cell's machinery through the process of protein turnover. A protein's lifespan is measured in terms of its half-life and covers a wide range. They can exist for minutes or years with an average lifespan of 1–2 days in mammalian cells. Abnormal or misfolded proteins are degraded more rapidly either due to being targeted for destruction or due to being unstable.

Like other biological macromolecules such as polysaccharides and nucleic acids, proteins are essential parts of organisms and participate in virtually every process within cells. Many proteins are enzymes that catalyse biochemical reactions and are vital to metabolism. Proteins also have structural or mechanical functions, such as actin and myosin in muscle and the proteins in the cytoskeleton, which form a system of scaffolding that maintains cell shape. Other proteins are important in cell signaling, immune responses, cell adhesion, and the cell cycle. In animals, proteins are needed in the diet to provide the essential amino acids that cannot be synthesized. Digestion breaks the proteins down for use in the metabolism.

Proteins may be purified from other cellular components using a variety of techniques such as ultracentrifugation, precipitation, electrophoresis, and chromatography; the advent of genetic engineering has made possible a number of methods to facilitate purification. Methods commonly used to study protein structure and function include immunohistochemistry, site-directed mutagenesis, X-ray crystallography, nuclear magnetic resonance and mass spectrometry.

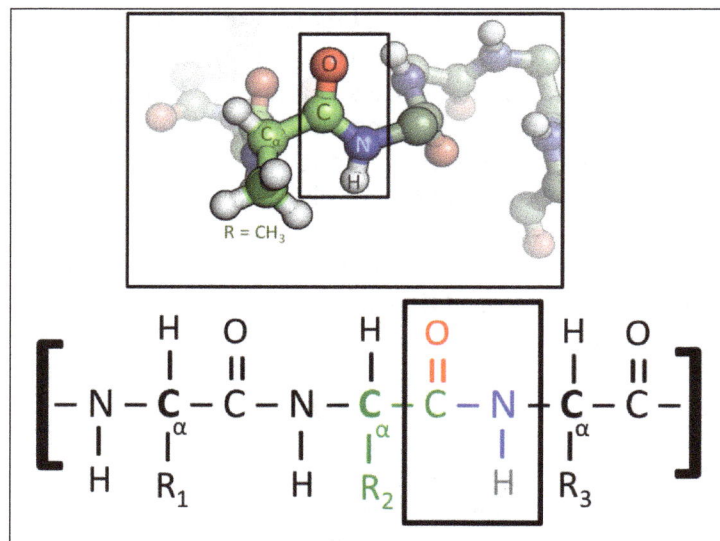

Chemical structure of the peptide bond (bottom) and the three-dimensional structure of a peptide bond between an alanine and an adjacent amino acid (top/inset). The bond itself is made of the CHON elements.

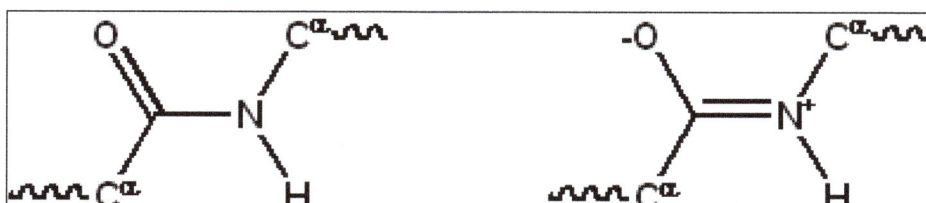

Resonance structures of the peptide bond that links individual amino acids to form a protein polymer.

Most proteins consist of linear polymers built from series of up to 20 different L-α- amino acids. All proteinogenic amino acids possess common structural features, including an α-carbon to which an amino group, a carboxyl group, and a variable side chain are bonded. Only proline differs from this basic structure as it contains an unusual ring to the N-end amine group, which forces the CO−NH amide moiety into a fixed conformation. The side chains of the standard amino acids, detailed in the list of standard amino acids, have a great variety of chemical structures and properties; it is the combined effect of all of the amino acid side chains in a protein that ultimately determines its three-dimensional structure and its chemical reactivity. The amino acids in a polypeptide chain are linked by peptide bonds. Once linked in the protein chain, an individual amino acid is called a *residue,* and the linked series of carbon, nitrogen, and oxygen atoms are known as the *main chain* or *protein backbone.*

The peptide bond has two resonance forms that contribute some double-bond character and inhibit rotation around its axis, so that the alpha carbons are roughly coplanar. The other two dihedral angles in the peptide bond determine the local shape assumed by the protein backbone. The end with a free amino group is known as the N-terminus or amino terminus, whereas the end of the protein with a free carboxyl group is known as the C-terminus or carboxy terminus (the sequence of the protein is written from N-terminus to C-terminus, from left to right).

The words *protein, polypeptide,* and *peptide* are a little ambiguous and can overlap in meaning. *Protein* is generally used to refer to the complete biological molecule in a stable conformation, whereas *peptide* is generally reserved for a short amino acid oligomers often lacking a stable three-dimensional structure. However, the boundary between the two is not well defined and usually lies near 20−30 residues. *Polypeptide* can refer to any single linear chain of amino acids, usually regardless of length, but often implies an absence of a defined conformation.

Interactions

Proteins can interact with many types of molecules, including with other proteins, with lipids, with carboyhydrates, and with DNA.

Abundance in Cells

It has been estimated that average-sized bacteria contain about 2 million proteins per cell (e.g. *E. coli* and *Staphylococcus aureus*). Smaller bacteria, such as *Mycoplasma* or *spirochetes* contain fewer molecules, on the order of 50,000 to 1 million. By contrast, eukaryotic cells are larger and thus contain much more protein. For instance, yeast cells have been estimated to contain about 50 million proteins and human cells on the order of 1 to 3 billion. The concentration of individual protein copies ranges from a few molecules per cell up to 20 million. Not all genes coding proteins are expressed in most cells and their number depends on, for example, cell type and external stimuli. For instance, of the 20,000 or so proteins encoded by the human genome, only 6,000 are detected in lymphoblastoid cells. Moreover, the number of proteins the genome encodes correlates well with the organism complexity. Eukaryotes have 15,000, bacteria have 3,200, archaea have 2,400, and viruses have 42 proteins on average coded in their respective genomes.

Synthesis

Biosynthesis

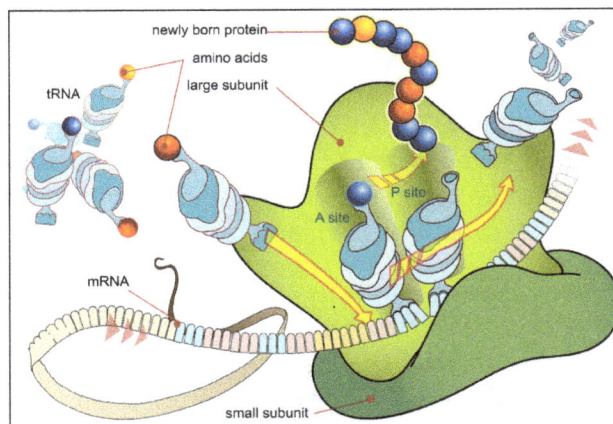

A ribosome produces a protein using mRNA as template.

Proteins are assembled from amino acids using information encoded in genes. Each protein has its own unique amino acid sequence that is specified by the nucleotide sequence of the gene encoding this protein. The genetic code is a set of three-nucleotide sets called codons and each three-nucleotide combination designates an amino acid, for example AUG (adenine-uracil-guanine) is the code for methionine. Because DNA contains four nucleotides, the total number of possible codons is 64; hence, there is some redundancy in the genetic code, with some amino acids specified by more than one codon. Genes encoded in DNA are first transcribed into pre-messenger RNA (mRNA) by proteins such as RNA polymerase. Most organisms then process the pre-mRNA (also known as a *primary transcript*) using various forms of Post-transcriptional modification to form the mature mRNA, which is then used as a template for protein synthesis by the ribosome. In prokaryotes the mRNA may either be used as soon as it is produced, or be bound by a ribosome after having moved away from the nucleoid. In contrast, eukaryotes make mRNA in the cell nucleus and then translocate it across the nuclear membrane into the cytoplasm, where protein synthesis then takes place. The rate of protein synthesis is higher in prokaryotes than eukaryotes and can reach up to 20 amino acids per second.

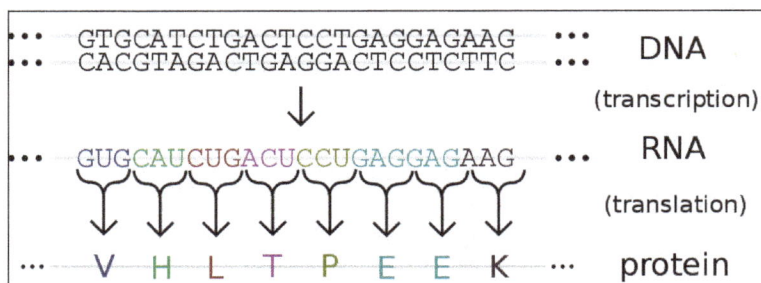

The DNA sequence of a gene encodes the amino acid sequence of a protein.

The process of synthesizing a protein from an mRNA template is known as translation. The mRNA is loaded onto the ribosome and is read three nucleotides at a time by matching each codon to its base pairing anticodon located on a transfer RNA molecule, which carries the amino acid corresponding to the codon it recognizes. The enzyme aminoacyl tRNA synthetase "charges" the tRNA molecules with the correct amino acids. The growing polypeptide is often termed the *nascent chain*. Proteins are always biosynthesized from N-terminus to C-terminus.

The size of a synthesized protein can be measured by the number of amino acids it contains and by its total molecular mass, which is normally reported in units of *daltons* (synonymous with atomic mass units), or the derivative unit kilodalton (kDa). The average size of a protein increases from Archaea to Bacteria to Eukaryote (283, 311, 438 residues and 31, 34, 49 kDa respecitvely) due to a bigger number of protein domains constituting proteins in higher organisms. For instance, yeast proteins are on average 466 amino acids long and 53 kDa in mass. The largest known proteins are the titins, a component of the muscle sarcomere, with a molecular mass of almost 3,000 kDa and a total length of almost 27,000 amino acids.

Chemical Synthesis

Short proteins can also be synthesized chemically by a family of methods known as peptide synthesis, which rely on organic synthesis techniques such as chemical ligation to produce peptides in high yield. Chemical synthesis allows for the introduction of non-natural amino acids into polypeptide chains, such as attachment of fluorescent probes to amino acid side chains. These methods are useful in laboratory biochemistry and cell biology, though generally not for commercial applications. Chemical synthesis is inefficient for polypeptides longer than about 300 amino acids, and the synthesized proteins may not readily assume their native tertiary structure. Most chemical synthesis methods proceed from C-terminus to N-terminus, opposite the biological reaction.

Structure

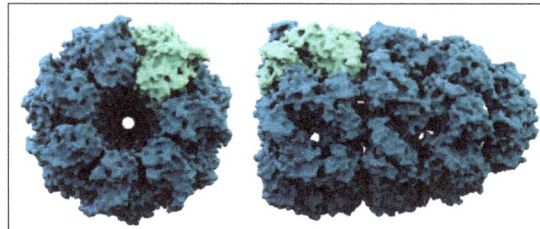

The crystal structure of the chaperonin, a huge protein complex. A single protein subunit is highlighted. Chaperonins assist protein folding.

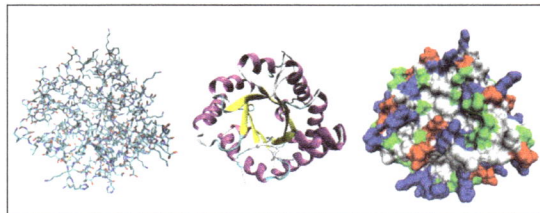

Three possible representations of the three-dimensional structure of the protein triose phosphate isomerase. Left: All-atom representation colored by atom type. Middle: Simplified representation illustrating the backbone conformation, colored by secondary structure. Right: Solvent-accessible surface representation colored by residue type (acidic residues red, basic residues blue, polar residues green, nonpolar residues white).

Most proteins fold into unique 3-dimensional structures. The shape into which a protein naturally folds is known as its native conformation. Although many proteins can fold unassisted, simply through the chemical properties of their amino acids, others require the aid of molecular chaperones to fold into their native states. Biochemists often refer to four distinct aspects of a protein's structure:

- Primary structure: The amino acid sequence. A protein is a polyamide.

- Secondary structure: Regularly repeating local structures stabilized by hydrogen bonds. The most common examples are the α-helix, β-sheet and turns. Because secondary structures are local, many regions of different secondary structure can be present in the same protein molecule.

- Tertiary structure: The overall shape of a single protein molecule; the spatial relationship of the secondary structures to one another. Tertiary structure is generally stabilized by non-local interactions, most commonly the formation of a hydrophobic core, but also through salt bridges, hydrogen bonds, disulfide bonds, and even posttranslational modifications. The term "tertiary structure" is often used as synonymous with the term fold. The tertiary structure is what controls the basic function of the protein.

- Quaternary structure: The structure formed by several protein molecules (polypeptide chains), usually called protein subunits in this context, which function as a single protein complex.

Proteins are not entirely rigid molecules. In addition to these levels of structure, proteins may shift between several related structures while they perform their functions. In the context of these functional rearrangements, these tertiary or quaternary structures are usually referred to as "conformations", and transitions between them are called *conformational changes*. Such changes are often induced by the binding of a substrate molecule to an enzyme's active site, or the physical region of the protein that participates in chemical catalysis. In solution proteins also undergo variation in structure through thermal vibration and the collision with other molecules.

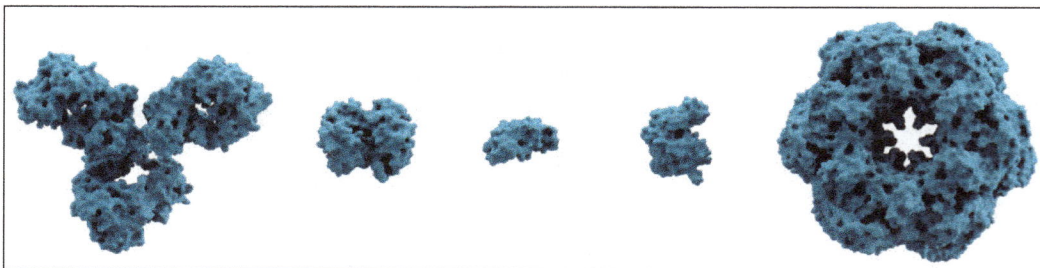

Molecular surface of several proteins showing their comparative sizes. From left to right are: immunoglobulin G (IgG, an antibody), hemoglobin, insulin (a hormone), adenylate kinase (an enzyme), and glutamine synthetase (an enzyme).

Proteins can be informally divided into three main classes, which correlate with typical tertiary structures: globular proteins, fibrous proteins, and membrane proteins. Almost all globular proteins are soluble and many are enzymes. Fibrous proteins are often structural, such as collagen, the major component of connective tissue, or keratin, the protein component of hair and nails.

Membrane proteins often serve as receptors or provide channels for polar or charged molecules to pass through the cell membrane.

A special case of intramolecular hydrogen bonds within proteins, poorly shielded from water attack and hence promoting their own dehydration, are called dehydrons.

Protein Domains

Many proteins are composed of several protein domains, i.e. segments of a protein that fold into distinct structural units. Domains usually also have specific functions, such as enzymatic activities (e.g. kinase) or they serve as binding modules (e.g. the SH3 domain binds to proline-rich sequences in other proteins).

Sequence Motif

Short amino acid sequences within proteins often act as recognition sites for other proteins. For instance, SH3 domains typically bind to short PxxP motifs (i.e. 2 prolines [P], separated by 2 unspecified amino acids [x], although the surrounding amino acids may determine the exact binding specificity). A large number of such motifs has been collected in the Eukaryotic Linear Motif (ELM) database.

Cellular Functions

Proteins are the chief actors within the cell, said to be carrying out the duties specified by the information encoded in genes. With the exception of certain types of RNA, most other biological molecules are relatively inert elements upon which proteins act. Proteins make up half the dry weight of an *Escherichia coli* cell, whereas other macromolecules such as DNA and RNA make up only 3% and 20%, respectively. The set of proteins expressed in a particular cell or cell type is known as its proteome.

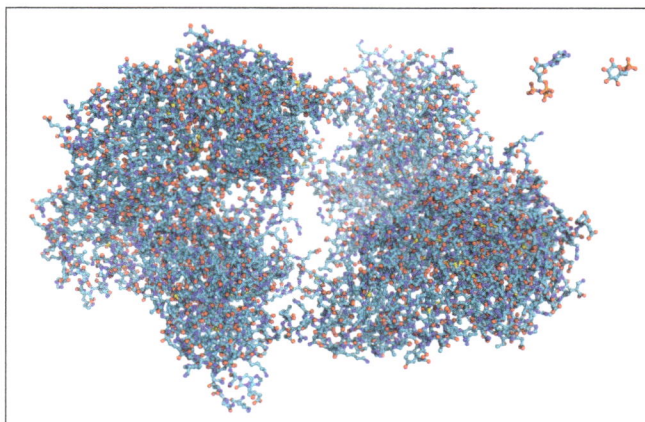

The enzyme hexokinase is shown as a conventional ball-and-stick molecular model. To scale in the top right-hand corner are two of its substrates, ATP and glucose.

The chief characteristic of proteins that also allows their diverse set of functions is their ability to bind other molecules specifically and tightly. The region of the protein responsible for binding another molecule is known as the binding site and is often a depression or "pocket" on the molecular surface. This binding ability is mediated by the tertiary structure of the protein, which defines the binding site pocket, and by the chemical properties of the surrounding amino acids' side chains. Protein binding can be extraordinarily tight and specific; for example, the ribonuclease inhibitor protein binds to human angiogenin with a sub-femtomolar dissociation constant ($<10^{-15}$ M) but does not bind at all to its amphibian homolog onconase (>1 M). Extremely minor chemical changes such as the addition of a single methyl group to a binding partner can sometimes suffice to nearly eliminate binding; for example, the aminoacyl tRNA synthetase specific to the amino acid valine discriminates against the very similar side chain of the amino acid isoleucine.

Proteins can bind to other proteins as well as to small-molecule substrates. When proteins bind specifically to other copies of the same molecule, they can oligomerize to form fibrils; this process occurs often in structural proteins that consist of globular monomers that self-associate to form rigid fibers. Protein–protein interactions also regulate enzymatic activity, control progression through the cell cycle, and allow the assembly of large protein complexes that carry out many closely related reactions with a common biological function. Proteins can also bind to, or even be integrated into, cell membranes. The ability of binding partners to induce conformational changes in proteins allows the construction of enormously complex signaling networks. As interactions between proteins are reversible, and depend heavily on the availability of different groups of partner proteins to form aggregates that are capable to carry out discrete sets of function, study of the interactions between specific proteins is a key to understand important aspects of cellular function, and ultimately the properties that distinguish particular cell types.

Enzymes

The best-known role of proteins in the cell is as enzymes, which catalyse chemical reactions. Enzymes are usually highly specific and accelerate only one or a few chemical reactions. Enzymes carry out most of the reactions involved in metabolism, as well as manipulating DNA in processes such as DNA replication, DNA repair, and transcription. Some enzymes act on other proteins to add or remove chemical groups in a process known as posttranslational modification. About 4,000 reactions are known to be catalysed by enzymes. The rate acceleration conferred by enzymatic catalysis is often enormous—as much as 10^{17}-fold increase in rate over the uncatalysed reaction in the case of orotate decarboxylase (78 million years without the enzyme, 18 milliseconds with the enzyme).

The molecules bound and acted upon by enzymes are called substrates. Although enzymes can consist of hundreds of amino acids, it is usually only a small fraction of the residues that come in contact with the substrate, and an even smaller fraction—three to four residues on average—that are directly involved in catalysis. The region of the enzyme that binds the substrate and contains the catalytic residues is known as the active site.

Dirigent proteins are members of a class of proteins that dictate the stereochemistry of a compound synthesized by other enzymes.

Cell Signaling and Ligand Binding

Many proteins are involved in the process of cell signaling and signal transduction. Some proteins, such as insulin, are extracellular proteins that transmit a signal from the cell in which they were synthesized to other cells in distant tissues. Others are membrane proteins that act as receptors whose main function is to bind a signaling molecule and induce a biochemical response in the cell. Many receptors have a binding site exposed on the cell surface and an effector domain within the cell, which may have enzymatic activity or may undergo a conformational change detected by other proteins within the cell.

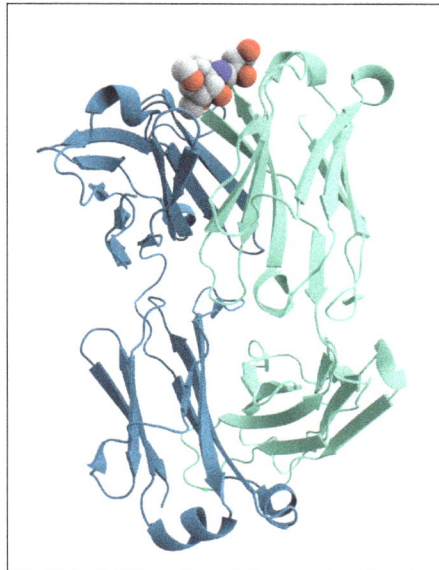

Ribbon diagram of a mouse antibody against cholera that binds a carbohydrate antigen.

Antibodies are protein components of an adaptive immune system whose main function is to bind antigens, or foreign substances in the body, and target them for destruction. Antibodies can be secreted into the extracellular environment or anchored in the membranes of specialized B cells known as plasma cells. Whereas enzymes are limited in their binding affinity for their substrates by the necessity of conducting their reaction, antibodies have no such constraints. An antibody's binding affinity to its target is extraordinarily high.

Many ligand transport proteins bind particular small biomolecules and transport them to other locations in the body of a multicellular organism. These proteins must have a high binding affinity when their ligand is present in high concentrations, but must also release the ligand when it is present at low concentrations in the target tissues. The canonical example of a ligand-binding protein is haemoglobin, which transports oxygen from the lungs to other organs and tissues in all vertebrates and has close homologs in every biological kingdom. Lectins are sugar-binding proteins which are highly specific for their sugar moieties. Lectins typically play a role in biological recognition phenomena involving cells and proteins. Receptors and hormones are highly specific binding proteins.

Transmembrane proteins can also serve as ligand transport proteins that alter the permeability of the cell membrane to small molecules and ions. The membrane alone has a hydrophobic core through which polar or charged molecules cannot diffuse. Membrane proteins contain internal channels that allow such molecules to enter and exit the cell. Many ion channel proteins are specialized to select for only a particular ion; for example, potassium and sodium channels often discriminate for only one of the two ions.

Structural Proteins

Structural proteins confer stiffness and rigidity to otherwise-fluid biological components. Most structural proteins are fibrous proteins; for example, collagen and elastin are critical components of connective tissue such as cartilage, and keratin is found in hard or filamentous structures such as hair, nails, feathers, hooves, and some animal shells. Some globular proteins can also play structural functions, for example, actin and tubulin are globular and soluble as monomers, but polymerize to form long, stiff fibers that make up the cytoskeleton, which allows the cell to maintain its shape and size.

Other proteins that serve structural functions are motor proteins such as myosin, kinesin, and dynein, which are capable of generating mechanical forces. These proteins are crucial for cellular motility of single celled organisms and the sperm of many multicellular organisms which reproduce sexually. They also generate the forces exerted by contracting muscles and play essential roles in intracellular transport.

The activities and structures of proteins may be examined *in vitro, in vivo, and in silico. In vitro* studies of purified proteins in controlled environments are useful for learning how a protein carries out its function: for example, enzyme kinetics studies explore the chemical mechanism of an enzyme's catalytic activity and its relative affinity for various possible substrate molecules. By contrast, *in vivo* experiments can provide information about the physiological role of a protein in the context of a cell or even a whole organism. *In silico* studies use computational methods to study proteins.

Protein Purification

To perform *in vitro* analysis, a protein must be purified away from other cellular components. This process usually begins with cell lysis, in which a cell's membrane is disrupted and its internal contents released into a solution known as a crude lysate. The resulting mixture can be purified using ultracentrifugation, which fractionates the various cellular components into fractions containing soluble proteins; membrane lipids and proteins; cellular organelles, and nucleic acids. Precipitation by a method known as salting out can concentrate the proteins from this lysate. Various types of chromatography are then used to isolate the protein or proteins of interest based on properties such as molecular weight, net charge and binding affinity. The level of purification can be monitored using various types of gel electrophoresis if the desired protein's molecular weight and isoelectric point are known, by spectroscopy if the protein has distinguishable spectroscopic features, or by enzyme assays if the protein has enzymatic activity. Additionally, proteins can be isolated according to their charge using electrofocusing.

For natural proteins, a series of purification steps may be necessary to obtain protein sufficiently pure for laboratory applications. To simplify this process, genetic engineering is often used to add chemical features to proteins that make them easier to purify without affecting their structure or activity. Here, a "tag" consisting of a specific amino acid sequence, often a series of histidine residues (a "His-tag"), is attached to one terminus of the protein. As a result, when the lysate is passed over a chromatography column containing nickel, the histidine residues ligate the nickel and attach to the column while the untagged components of the lysate pass unimpeded. A number of different tags have been developed to help researchers purify specific proteins from complex mixtures.

Cellular Localization

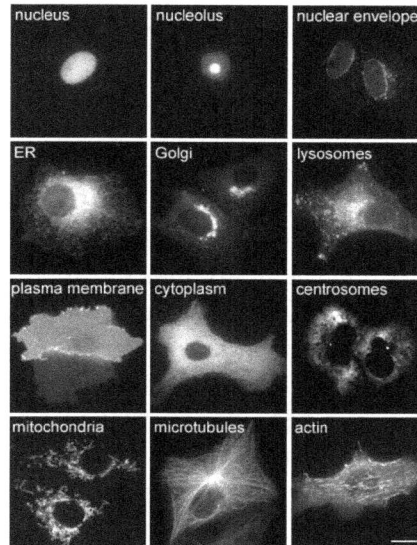

Proteins in different cellular compartments and structures tagged with green fluorescent protein (here, white).

The study of proteins *in vivo* is often concerned with the synthesis and localization of the protein within the cell. Although many intracellular proteins are synthesized in the cytoplasm and membrane-bound or secreted proteins in the endoplasmic reticulum, the specifics of how proteins are targeted to specific organelles or cellular structures is often unclear. A useful technique for assessing cellular localization uses genetic engineering to express in a cell a fusion protein or chimera consisting of the natural protein of interest linked to a "reporter" such as green fluorescent protein (GFP). The fused protein's position within the cell can be cleanly and efficiently visualized using microscopy, as shown in the figure opposite.

Other methods for elucidating the cellular location of proteins requires the use of known compartmental markers for regions such as the ER, the Golgi, lysosomes or vacuoles, mitochondria, chloroplasts, plasma membrane, etc. With the use of fluorescently tagged versions of these markers or of antibodies to known markers, it becomes much simpler to identify the localization of a protein of interest. For example, indirect immunofluorescence will allow for fluorescence colocalization and demonstration of location. Fluorescent dyes are used to label cellular compartments for a similar purpose.

Other possibilities exist, as well. For example, immunohistochemistry usually utilizes an antibody to one or more proteins of interest that are conjugated to enzymes yielding either luminescent or chromogenic signals that can be compared between samples, allowing for localization information. Another applicable technique is cofractionation in sucrose (or other material) gradients using isopycnic centrifugation. While this technique does not prove colocalization of a compartment of known density and the protein of interest, it does increase the likelihood, and is more amenable to large-scale studies.

Finally, the gold-standard method of cellular localization is immunoelectron microscopy. This technique also uses an antibody to the protein of interest, along with classical electron microscopy techniques. The sample is prepared for normal electron microscopic examination, and then treated with an antibody to the protein of interest that is conjugated to an extremely electro-dense material, usually gold. This allows for the localization of both ultrastructural details as well as the protein of interest.

Through another genetic engineering application known as site-directed mutagenesis, researchers can alter the protein sequence and hence its structure, cellular localization, and susceptibility to regulation. This technique even allows the incorporation of unnatural amino acids into proteins, using modified tRNAs, and may allow the rational design of new proteins with novel properties.

Proteomics

The total complement of proteins present at a time in a cell or cell type is known as its proteome, and the study of such large-scale data sets defines the field of proteomics, named by analogy to the related field of genomics. Key experimental techniques in proteomics include 2D electrophoresis, which allows the separation of a large number of proteins, mass spectrometry, which allows rapid high-throughput identification of proteins and sequencing of peptides (most often after in-gel digestion), protein microarrays, which allow the detection of the relative levels of a large number of proteins present in a cell, and two-hybrid screening, which allows the systematic exploration of protein–protein interactions. The total complement of biologically possible such interactions is known as the interactome. A systematic attempt to determine the structures of proteins representing every possible fold is known as structural genomics.

Bioinformatics

A vast array of computational methods have been developed to analyze the structure, function, and evolution of proteins. The development of such tools has been driven by the large amount of genomic and proteomic data available for a variety of organisms, including the human genome. It is simply impossible to study all proteins experimentally, hence only a few are subjected to laboratory experiments while computational tools are used to extrapolate to similar proteins. Such homologous proteins can be efficiently identified in distantly related organisms by sequence alignment. Genome and gene sequences can be searched by a variety of tools for certain properties. Sequence profiling tools can find restriction enzyme sites, open reading frames in nucleotide sequences, and predict secondary structures. Phylogenetic trees can be constructed and evolutionary hypotheses developed using special software like ClustalW regarding the ancestry of modern organisms and the genes they express. The field of bioinformatics is now indispensable for the analysis of genes and proteins.

Structure Determination

Discovering the tertiary structure of a protein, or the quaternary structure of its complexes, can provide important clues about how the protein performs its function and how it can be affected, i.e. in drug design. As proteins are too small to be seen under a light microscope, other methods have to be employed to determine their structure. Common experimental methods include X-ray crystallography and NMR spectroscopy, both of which can produce structural information at atomic resolution. However, NMR experiments are able to provide information from which a subset of distances between pairs of atoms can be estimated, and the final possible conformations for a protein are determined by solving a distance geometry problem. Dual polarisation interferometry is a quantitative analytical method for measuring the overall protein conformation and conformational changes due to interactions or other stimulus. Circular dichroism is another laboratory technique for determining internal β-sheet / α-helical composition of proteins. Cryo-electron microscopy is used to produce lower-resolution structural information about very large protein complexes, including assembled viruses; a variant known as electron crystallography can also produce high-resolution information in some cases, especially for two-dimensional crystals of membrane proteins. Solved structures are usually deposited in the Protein Data Bank (PDB), a freely available resource from which structural data about thousands of proteins can be obtained in the form of Cartesian coordinates for each atom in the protein.

Many more gene sequences are known than protein structures. Further, the set of solved structures is biased toward proteins that can be easily subjected to the conditions required in X-ray crystallography, one of the major structure determination methods. In particular, globular proteins are comparatively easy to crystallize in preparation for X-ray crystallography. Membrane proteins and large protein complexes, by contrast, are difficult to crystallize and are underrepresented in the PDB. Structural genomics initiatives have attempted to remedy these deficiencies by systematically solving representative structures of major fold classes. Protein structure prediction methods attempt to provide a means of generating a plausible structure for proteins whose structures have not been experimentally determined.

Structure Prediction and Simulation

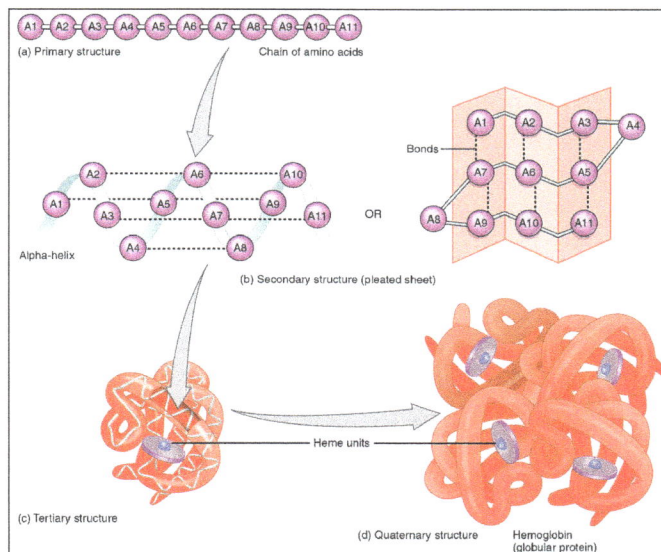

Constituent amino-acids can be analyzed to predict secondary, tertiary and quaternary protein structure, in this case hemoglobin containing heme units.

Complementary to the field of structural genomics, *protein structure prediction* develops efficient mathematical models of proteins to computationally predict the molecular formations in theory, instead of detecting structures with laboratory observation. The most successful type of structure prediction, known as homology modeling, relies on the existence of a "template" structure with sequence similarity to the protein being modeled; structural genomics' goal is to provide sufficient representation in solved structures to model most of those that remain. Although producing accurate models remains a challenge when only distantly related template structures are available, it has been suggested that sequence alignment is the bottleneck in this process, as quite accurate models can be produced if a "perfect" sequence alignment is known. Many structure prediction methods have served to inform the emerging field of protein engineering, in which novel protein folds have already been designed. A more complex computational problem is the prediction of intermolecular interactions, such as in molecular docking and protein–protein interaction prediction.

Mathematical models to simulate dynamic processes of protein folding and binding involve molecular mechanics, in particular, molecular dynamics. Monte Carlo techniques facilitate the computations, which exploit advances in parallel and distributed computing (for example, the Folding@ home project which performs molecular modeling on GPUs). *In silico* simulations discovered the folding of small α-helical protein domains such as the villin headpiece and the HIV accessory protein. Hybrid methods combining standard molecular dynamics with quantum mechanical mathematics explored the electronic states of rhodopsins.

Protein Disorder and Unstructure Prediction

Many proteins (in Eucaryota ~33%) contain large unstructured but biologically functional segments and can be classified as intrinsically disordered proteins. Predicting and analysing protein disorder is, therefore, an important part of protein structure characterisation.

Nutrition

Most microorganisms and plants can biosynthesize all 20 standard amino acids, while animals (including humans) must obtain some of the amino acids from the diet. The amino acids that an organism cannot synthesize on its own are referred to as essential amino acids. Key enzymes that synthesize certain amino acids are not present in animals—such as aspartokinase, which catalyses the first step in the synthesis of lysine, methionine, and threonine from aspartate. If amino acids are present in the environment, microorganisms can conserve energy by taking up the amino acids from their surroundings and downregulating their biosynthetic pathways.

In animals, amino acids are obtained through the consumption of foods containing protein. Ingested proteins are then broken down into amino acids through digestion, which typically involves denaturation of the protein through exposure to acid and hydrolysis by enzymes called proteases. Some ingested amino acids are used for protein biosynthesis, while others are converted to glucose through gluconeogenesis, or fed into the citric acid cycle. This use of protein as a fuel is particularly important under starvation conditions as it allows the body's own proteins to be used to support life, particularly those found in muscle.

In animals such as dogs and cats, protein maintains the health and quality of the skin by promoting hair follicle growth and keratinization, and thus reducing the likelihood of skin problems producing malodours. Poor-quality proteins also have a role regarding gastrointestinal health, increasing the potential for flatulence and odorous compounds in dogs because when proteins reach the colon in an undigested state, they are fermented producing hydrogen sulfide gas, indole, and skatole. Dogs and cats digest animal proteins better than those from plants but products of low-quality animal origin are poorly digested, including skin, feathers, and connective tissue.

DNA

Deoxyribonucleic acid is a molecule composed of two chains that coil around each other to form a double helix carrying genetic instructions for the development, functioning, growth and reproduction of all known organisms and many viruses. DNA and ribonucleic acid (RNA) are nucleic acids;

alongside proteins, lipids and complex carbohydrates (polysaccharides), nucleic acids are one of the four major types of macromolecules that are essential for all known forms of life.

The two DNA strands are also known as polynucleotides as they are composed of simpler monomeric units called nucleotides. Each nucleotide is composed of one of four nitrogen-containing nucleobases (cytosine [C], guanine [G], adenine [A] or thymine [T]), a sugar called deoxyribose, and a phosphate group. The nucleotides are joined to one another in a chain by covalent bonds between the sugar of one nucleotide and the phosphate of the next, resulting in an alternating sugar-phosphate backbone. The nitrogenous bases of the two separate polynucleotide strands are bound together, according to base pairing rules (A with T and C with G), with hydrogen bonds to make double-stranded DNA. The complementary nitrogenous bases are divided into two groups, pyrimidines and purines. In DNA, the pyrimidines are thymine and cytosine; the purines are adenine and guanine.

Both strands of double-stranded DNA store the same biological information. This information is replicated as and when the two strands separate. A large part of DNA (more than 98% for humans) is non-coding, meaning that these sections do not serve as patterns for protein sequences. The two strands of DNA run in opposite directions to each other and are thus antiparallel. Attached to each sugar is one of four types of nucleobases (informally, *bases*). It is the sequence of these four nucleobases along the backbone that encodes genetic information. RNA strands are created using DNA strands as a template in a process called transcription. Under the genetic code, these RNA strands specify the sequence of amino acids within proteins in a process called translation.

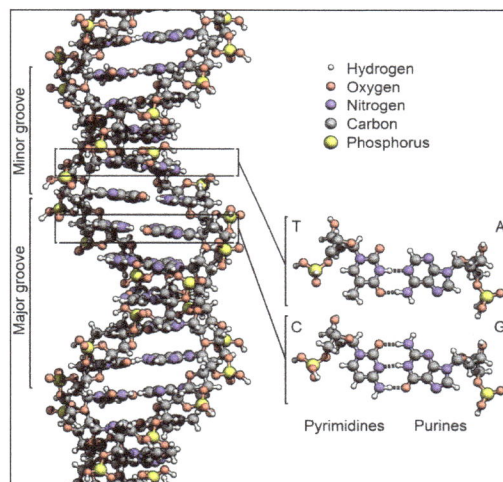

The structure of the DNA double helix. The atoms in the structure are colour-coded by element and the detailed structures of two base pairs are shown in the bottom right.

Within eukaryotic cells, DNA is organized into long structures called chromosomes. Before typical cell division, these chromosomes are duplicated in the process of DNA replication, providing a complete set of chromosomes for each daughter cell. Eukaryotic organisms (animals, plants, fungi and protists) store most of their DNA inside the cell nucleus as nuclear DNA, and some in the mitochondria as mitochondrial DNA or in chloroplasts as chloroplast DNA. In contrast, prokaryotes (bacteria and archaea) store their DNA only in the cytoplasm, in circular chromosomes. Within eukaryotic chromosomes, chromatin proteins, such as histones, compact and organize DNA. These compacting structures guide the interactions between DNA and other proteins, helping control which parts of the DNA are transcribed.

DNA was first isolated by Friedrich Miescher in 1869. Its molecular structure was first identified by Francis Crick and James Watson at the Cavendish Laboratory within the University of Cambridge in 1953, whose model-building efforts were guided by X-ray diffraction data acquired by Raymond Gosling, who was a post-graduate student of Rosalind Franklin. DNA is used by researchers as a molecular tool to explore physical laws and theories, such as the ergodic theorem and the theory of elasticity. The unique material properties of DNA have made it an attractive molecule for material scientists and engineers interested in micro- and nano-fabrication. Among notable advances in this field are DNA origami and DNA-based hybrid materials.

Properties

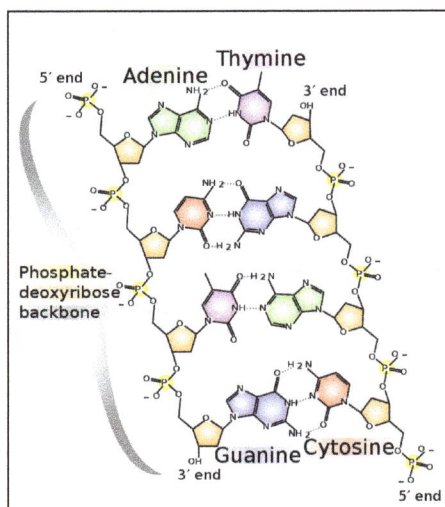

Chemical structure of DNA; hydrogen bonds shown as dotted lines.

DNA is a long polymer made from repeating units called nucleotides. The structure of DNA is dynamic along its length, being capable of coiling into tight loops and other shapes. In all species it is composed of two helical chains, bound to each other by hydrogen bonds. Both chains are coiled around the same axis, and have the same pitch of 34 angstroms (Å) (3.4 nanometres). The pair of chains has a radius of 10 angstroms (1.0 nanometre). According to another study, when measured in a different solution, the DNA chain measured 22 to 26 angstroms wide (2.2 to 2.6 nanometres), and one nucleotide unit measured 3.3 Å (0.33 nm) long. Although each individual nucleotide is very small, a DNA polymer can be very large and contain hundreds of millions, such as in chromosome 1. Chromosome 1 is the largest human chromosome with approximately 220 million base pairs, and would be 85 mm long if straightened.

DNA does not usually exist as a single strand, but instead as a pair of strands that are held tightly together. These two long strands coil around each other, in the shape of a double helix. The nucleotide contains both a segment of the backbone of the molecule (which holds the chain together) and a nucleobase (which interacts with the other DNA strand in the helix). A nucleobase linked to a sugar is called a nucleoside, and a base linked to a sugar and to one or more phosphate groups is called a nucleotide. A biopolymer comprising multiple linked nucleotides (as in DNA) is called a polynucleotide.

The backbone of the DNA strand is made from alternating phosphate and sugar residues. The sugar in DNA is 2-deoxyribose, which is a pentose (five-carbon) sugar. The sugars are joined together by phosphate groups that form phosphodiester bonds between the third and fifth carbon atoms of adjacent sugar rings. These are known as the 3′-end (three prime end), and 5′-end (five prime end) carbons, the prime symbol being used to distinguish these carbon atoms from those of the base to which the deoxyribose forms a glycosidic bond. When imagining DNA, each phosphoryl is normally considered to "belong" to the nucleotide whose 5′ carbon forms a bond therewith. Any DNA strand therefore normally has one end at which there is a phosphoryl attached to the 5′ carbon of a ribose (the 5′ phosphoryl) and another end at which there is a free hydroxyl attached to the 3′ carbon of a ribose (the 3′ hydroxyl). The orientation of the 3′ and 5′ carbons along the sugar-phosphate backbone confers directionality (sometimes called polarity) to each DNA strand. In a nucleic acid double helix, the direction of the nucleotides in one strand is opposite to their direction in the other strand: the strands are antiparallel. The asymmetric ends of DNA strands are said to have a directionality of five prime end (5′), and three prime end (3′), with the 5′ end having a terminal phosphate group and the 3′ end a terminal hydroxyl group. One major difference between DNA and RNA is the sugar, with the 2-deoxyribose in DNA being replaced by the alternative pentose sugar ribose in RNA.

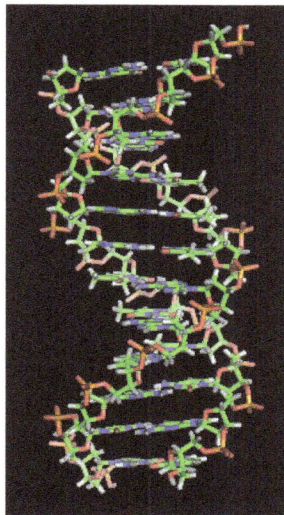

A section of DNA. The bases lie horizontally between the two spiraling strands.

The DNA double helix is stabilized primarily by two forces: hydrogen bonds between nucleotides and base-stacking interactions among aromatic nucleobases. In the cytosol of the cell, the conjugated pi bonds of nucleotide bases align perpendicular to the axis of the DNA molecule, minimizing their interaction with the solvation shell. The four bases found in DNA are adenine (A), cytosine (C), guanine (G) and thymine (T). These four bases are attached to the sugar-phosphate to form the complete nucleotide, as shown for adenosine monophosphate. Adenine pairs with thymine and guanine pairs with cytosine, forming A-T and G-C base pairs.

Nucleobase Classification

The nucleobases are classified into two types: the purines, A and G, which are fused five- and six-membered heterocyclic compounds, and the pyrimidines, the six-membered rings C and T. A fifth pyrimidine nucleobase, uracil (U), usually takes the place of thymine in RNA and differs from thymine by lacking a methyl group on its ring. In addition to RNA and DNA, many artificial nucleic acid analogues have been created to study the properties of nucleic acids, or for use in biotechnology.

Non-canonical Bases

Uracil is not usually found in DNA, occurring only as a breakdown product of cytosine. However, in several bacteriophages, such as *Bacillus subtilis* phages PBS1 and PBS2 and *Yersinia* phage piR1-37, thymine has been replaced by uracil. Another phage—Staphylococcal phage S6—has been identified with a genome where thymine has been replaced by uracil.

Uracil is also found in the DNA of *Plasmodium falciparum* It is present is relatively small amounts (7–10 uracil residues per million bases).

5-hydroxymethyldeoxyuridine,(hm5dU) is also known to replace thymidine in several genomes including the *Bacillus phages* SPO1, φe, SP8, H1, 2C and SP82.

Base J (beta-d-glucopyranosyloxymethyluracil), a modified form of uracil, is also found in several organisms: the flagellates *Diplonema* and *Euglena*, and all the kinetoplastid genera. Biosynthesis of J occurs in two steps: in the first step, a specific thymidine in DNA is converted into hydroxymethyldeoxyuridine; in the second, HOMedU is glycosylated to form J. Proteins that bind specifically to this base have been identified. These proteins appear to be distant relatives of the Tet1 oncogene that is involved in the pathogenesis of acute myeloid leukemia. J appears to act as a termination signal for RNA polymerase II.

In 1976, the S-2La bacteriophage, which infects species of the genus *Synechocystis*, was found to have all the adenosine bases within its genome replaced by 2,6-diaminopurine. In 2016 deoxyarchaeosine was found to be present in the genomes of several bacteria and the *Escherichia* phage 9g.

Modified bases also occur in DNA. The first of these recognised was 5-methylcytosine, which was found in the genome of *Mycobacterium tuberculosis* in 1925. The complete replacement of cytosine by 5-glycosylhydroxymethylcytosine in T even phages (T2, T4 and T6) was observed in 1953. In the genomes of Xanthomonas oryzae bacteriophage Xp12 and halovirus FH the full complement of cystosine has been replaced by 5-methylcytosine. 6N-methyladenine was discovered to be present in DNA in 1955. N6-carbamoyl-methyladenine was described in 1975. 7-Methylguanine was described in 1976. N4-methylcytosine in DNA was described in 1983. In 1985 5-hydroxycytosine was found in the genomes of the Rhizobium phages RL38JI and N17. α-putrescinylthymine occurs in both the genomes of the *Delftia* phage ΦW-14 and the *Bacillus* phage SP10. α-glutamylthymidine is found in the Bacillus phage SP01 and 5-dihydroxypentyluracil is found in the Bacillus phage SP15.

The reason for the presence of these non canonical bases in DNA is not known. It seems likely that at least part of the reason for their presence in bacterial viruses (phages) is to avoid the restriction enzymes present in bacteria. This enzyme system acts at least in part as a molecular immune system protecting bacteria from infection by viruses.

This does not appear to be the entire story. Four modifications to the cytosine residues in human DNA have been reported. These modifications are the addition of methyl (CH_3)-, hydroxymethyl (CH_2OH)-, formyl (CHO)- and carboxyl (COOH)- groups. These modifications are thought to have regulatory functions.

Uracil is found in the centromeric regions of at least two human chromosomes (chromosome 6 and chromosome 11).

Listing of Non Canonical bases found in DNA

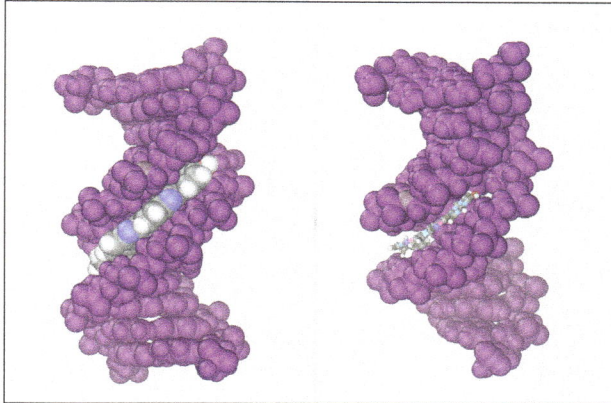

DNA major and minor grooves. The latter is a binding site for the Hoechst stain dye 33258.

Seventeen non canonical bases are known to occur in DNA. Most of these are modifications of the canonical bases plus uracil.

- Modified Adenosine:
  - N6-carbamoyl-methyladenine
  - N6-methyadenine
- Modified Guanine:
  - 7-Methylguanine
- Modified Cytosine:
  - N4-Methylcytosine
  - 5-Carboxylcytosine
  - 5-Formylcytosine
  - 5-Glycosylhydroxymethylcytosine
  - 5-Hydroxycytosine
  - 5-Methylcytosine
- Modified Thymidine:
  - α-Glutamythymidine
  - α-Putrescinylthymine
- Uracil and modifications:
  - Base J
  - Uracil

     ∘   5-Dihydroxypentauracil

     ∘   5-Hydroxymethyldeoxyuracil

- Others:

     ∘   Deoxyarchaeosine

     ∘   2,6-Diaminopurine

Grooves

Twin helical strands form the DNA backbone. Another double helix may be found tracing the spaces, or grooves, between the strands. These voids are adjacent to the base pairs and may provide a binding site. As the strands are not symmetrically located with respect to each other, the grooves are unequally sized. One groove, the major groove, is 22 angstroms (Å) wide and the other, the minor groove, is 12 Å wide. The width of the major groove means that the edges of the bases are more accessible in the major groove than in the minor groove. As a result, proteins such as transcription factors that can bind to specific sequences in double-stranded DNA usually make contact with the sides of the bases exposed in the major groove. This situation varies in unusual conformations of DNA within the cell, but the major and minor grooves are always named to reflect the differences in size that would be seen if the DNA is twisted back into the ordinary B form.

Base Pairing

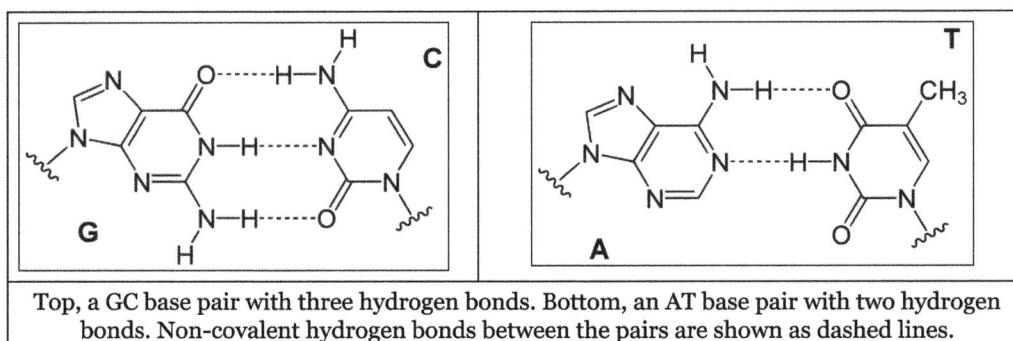

Top, a GC base pair with three hydrogen bonds. Bottom, an AT base pair with two hydrogen bonds. Non-covalent hydrogen bonds between the pairs are shown as dashed lines.

In a DNA double helix, each type of nucleobase on one strand bonds with just one type of nucleobase on the other strand. This is called complementary base pairing. Here, purines form hydrogen bonds to pyrimidines, with adenine bonding only to thymine in two hydrogen bonds, and cytosine bonding only to guanine in three hydrogen bonds. This arrangement of two nucleotides binding together across the double helix is called a Watson-Crick base pair. Another type of base pairing is Hoogsteen base pairing where two hydrogen bonds form between guanine and cytosine. As hydrogen bonds are not covalent, they can be broken and rejoined relatively easily. The two strands of DNA in a double helix can thus be pulled apart like a zipper, either by a mechanical force or high temperature. As a result of this base pair complementarity, all the information in the double-stranded sequence of a DNA helix is duplicated on each strand, which is vital in DNA replication. This reversible and specific interaction between complementary base pairs is critical for all the functions of DNA in organisms.

The two types of base pairs form different numbers of hydrogen bonds, AT forming two hydrogen bonds, and GC forming three hydrogen bonds. DNA with high GC-content is more stable than DNA with low GC-content.

Most DNA molecules are actually two polymer strands, bound together in a helical fashion by non-covalent bonds; this double-stranded (dsDNA) structure is maintained largely by the intrastrand base stacking interactions, which are strongest for G,C stacks. The two strands can come apart—a process known as melting—to form two single-stranded DNA (ssDNA) molecules. Melting occurs at high temperature, low salt and high pH (low pH also melts DNA, but since DNA is unstable due to acid depurination, low pH is rarely used).

The stability of the dsDNA form depends not only on the GC-content (% G,C basepairs) but also on sequence (since stacking is sequence specific) and also length (longer molecules are more stable). The stability can be measured in various ways; a common way is the "melting temperature", which is the temperature at which 50% of the ds molecules are converted to ss molecules; melting temperature is dependent on ionic strength and the concentration of DNA. As a result, it is both the percentage of GC base pairs and the overall length of a DNA double helix that determines the strength of the association between the two strands of DNA. Long DNA helices with a high GC-content have stronger-interacting strands, while short helices with high AT content have weaker-interacting strands. In biology, parts of the DNA double helix that need to separate easily, such as the TATAAT Pribnow box in some promoters, tend to have a high AT content, making the strands easier to pull apart.

In the laboratory, the strength of this interaction can be measured by finding the temperature necessary to break the hydrogen bonds, their melting temperature (also called T_m value). When all the base pairs in a DNA double helix melt, the strands separate and exist in solution as two entirely independent molecules. These single-stranded DNA molecules have no single common shape, but some conformations are more stable than others.

Sense and Antisense

A DNA sequence is called a "sense" sequence if it is the same as that of a messenger RNA copy that is translated into protein. The sequence on the opposite strand is called the "antisense" sequence. Both sense and antisense sequences can exist on different parts of the same strand of DNA (i.e. both strands can contain both sense and antisense sequences). In both prokaryotes and eukaryotes, antisense RNA sequences are produced, but the functions of these RNAs are not entirely clear. One proposal is that antisense RNAs are involved in regulating gene expression through RNA-RNA base pairing.

A few DNA sequences in prokaryotes and eukaryotes, and more in plasmids and viruses, blur the distinction between sense and antisense strands by having overlapping genes. In these cases, some DNA sequences do double duty, encoding one protein when read along one strand, and a second protein when read in the opposite direction along the other strand. In bacteria, this overlap may be involved in the regulation of gene transcription, while in viruses, overlapping genes increase the amount of information that can be encoded within the small viral genome.

Supercoiling

DNA can be twisted like a rope in a process called DNA supercoiling. With DNA in its "relaxed" state, a strand usually circles the axis of the double helix once every 10.4 base pairs, but if the DNA is twisted the strands become more tightly or more loosely wound. If the DNA is twisted in the direction of the helix, this is positive supercoiling, and the bases are held more tightly together. If they are twisted in the opposite direction, this is negative supercoiling, and the bases come apart more easily. In nature, most DNA has slight negative supercoiling that is introduced by enzymes called topoisomerases. These enzymes are also needed to relieve the twisting stresses introduced into DNA strands during processes such as transcription and DNA replication.

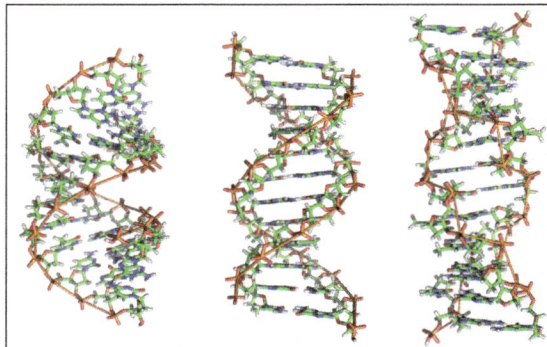

From left to right, the structures of A, B and Z DNA.

Alternative DNA Structures

DNA exists in many possible conformations that include A-DNA, B-DNA, and Z-DNA forms, although, only B-DNA and Z-DNA have been directly observed in functional organisms. The conformation that DNA adopts depends on the hydration level, DNA sequence, the amount and direction of supercoiling, chemical modifications of the bases, the type and concentration of metal ions, and the presence of polyamines in solution.

The first published reports of A-DNA X-ray diffraction patterns—and also B-DNA—used analyses based on Patterson transforms that provided only a limited amount of structural information for oriented fibers of DNA. An alternative analysis was then proposed by Wilkins *et al.*, in 1953, for the *in vivo* B-DNA X-ray diffraction-scattering patterns of highly hydrated DNA fibers in terms of squares of Bessel functions. In the same journal, James Watson and Francis Crick presented their molecular modeling analysis of the DNA X-ray diffraction patterns to suggest that the structure was a double-helix.

Although the *B-DNA form* is most common under the conditions found in cells, it is not a well-defined conformation but a family of related DNA conformations that occur at the high hydration levels present in cells. Their corresponding X-ray diffraction and scattering patterns are characteristic of molecular paracrystals with a significant degree of disorder.

Compared to B-DNA, the A-DNA form is a wider right-handed spiral, with a shallow, wide minor groove and a narrower, deeper major groove. The A form occurs under non-physiological conditions in partly dehydrated samples of DNA, while in the cell it may be produced in hybrid pairings of DNA and RNA strands, and in enzyme-DNA complexes. Segments of DNA where the bases have been chemically modified by methylation may undergo a larger change in conformation and adopt the Z form. Here, the strands turn about the helical axis in a left-handed spiral, the opposite of the more common B form. These unusual structures can be recognized by specific Z-DNA binding proteins and may be involved in the regulation of transcription.

Alternative DNA Chemistry

For many years, exobiologists have proposed the existence of a shadow biosphere, a postulated microbial biosphere of Earth that uses radically different biochemical and molecular processes than currently known life. One of the proposals was the existence of lifeforms that use arsenic instead of phosphorus in DNA. A report in 2010 of the possibility in the bacterium GFAJ-1, was announced, though the research was disputed, and evidence suggests the bacterium actively prevents the incorporation of arsenic into the DNA backbone and other biomolecules.

Quadruplex Structures

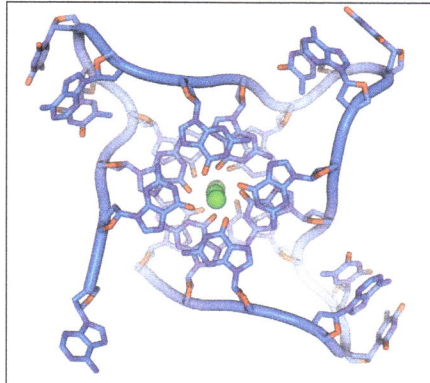

DNA quadruplex formed by telomere repeats. The looped conformation of the DNA backbone is very different from the typical DNA helix. The green spheres in the center represent potassium ions.

At the ends of the linear chromosomes are specialized regions of DNA called telomeres. The main function of these regions is to allow the cell to replicate chromosome ends using the enzyme telomerase, as the enzymes that normally replicate DNA cannot copy the extreme 3′ ends of chromosomes. These specialized chromosome caps also help protect the DNA ends, and stop the DNA repair systems in the cell from treating them as damage to be corrected. In human cells, telomeres are usually lengths of single-stranded DNA containing several thousand repeats of a simple TTAGGG sequence.

These guanine-rich sequences may stabilize chromosome ends by forming structures of stacked sets of four-base units, rather than the usual base pairs found in other DNA molecules. Here, four guanine bases form a flat plate and these flat four-base units then stack on top of each other, to form a stable G-quadruplex structure. These structures are stabilized by hydrogen bonding between the edges of the bases and chelation of a metal ion in the centre of each four-base unit. Other structures can also be formed, with the central set of four bases coming from either a single strand folded around the bases, or several different parallel strands, each contributing one base to the central structure.

In addition to these stacked structures, telomeres also form large loop structures called telomere loops, or T-loops. Here, the single-stranded DNA curls around in a long circle stabilized by telomere-binding proteins. At the very end of the T-loop, the single-stranded telomere DNA is held onto a region of double-stranded DNA by the telomere strand disrupting the double-helical DNA and base pairing to one of the two strands. This triple-stranded structure is called a displacement loop or D-loop.

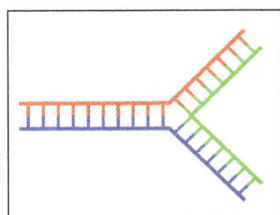

Single branch   Multiple branches

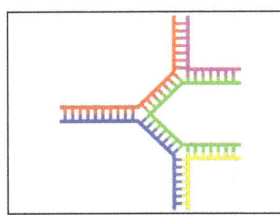

Branched DNA

In DNA, fraying occurs when non-complementary regions exist at the end of an otherwise complementary double-strand of DNA. However, branched DNA can occur if a third strand of DNA is introduced and contains adjoining regions able to hybridize with the frayed regions of the pre-existing double-strand. Although the simplest example of branched DNA involves only three strands of DNA, complexes involving additional strands and multiple branches are also possible. Branched DNA can be used in nanotechnology to construct geometric shapes.

Artificial Bases

Several artificial nucleobases have been synthesized, and successfully incorporated in the eight-base DNA analogue named Hachimoji DNA. Dubbed S, B, P, and Z, these artificial bases are capable of bonding with each other in a predictable way (S–B and P–Z), maintain the double helix structure of DNA, and be transcribed to RNA. Their existence implies that there is nothing special about the four natural nucleobases that evolved on Earth.

Chemical Modifications and Altered DNA Packaging

Structure of cytosine with and without the 5-methyl group. Deamination converts 5-methylcytosine into thymine.

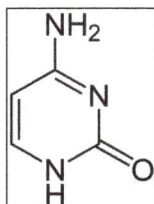

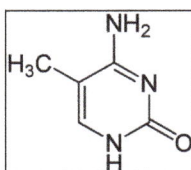

Cytosine   5-methylcytosine   Thymine

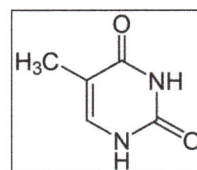

Base Modifications and DNA Packaging

The expression of genes is influenced by how the DNA is packaged in chromosomes, in a structure called chromatin. Base modifications can be involved in packaging, with regions that have low or no gene expression usually containing high levels of methylation of cytosine bases. DNA packaging and its influence on gene expression can also occur by covalent modifications of the histone protein core around which DNA is wrapped in the chromatin structure or else by remodeling carried out by chromatin remodeling complexes. There is, further, crosstalk between

DNA methylation and histone modification, so they can coordinately affect chromatin and gene expression.

For one example, cytosine methylation produces 5-methylcytosine, which is important for X-inactivation of chromosomes. The average level of methylation varies between organisms—the worm *Caenorhabditis elegans* lacks cytosine methylation, while vertebrates have higher levels, with up to 1% of their DNA containing 5-methylcytosine. Despite the importance of 5-methylcytosine, it can deaminate to leave a thymine base, so methylated cytosines are particularly prone to mutations. Other base modifications include adenine methylation in bacteria, the presence of 5-hydroxymethylcytosine in the brain, and the glycosylation of uracil to produce the "J-base" in kinetoplastids.

Damage

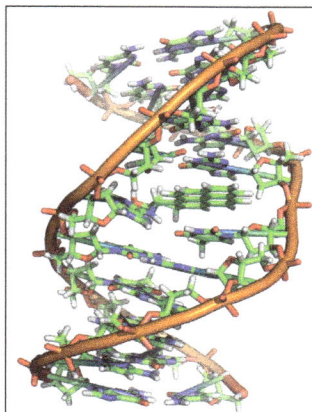

A covalent adduct between a metabolically activated form of benzo[*a*]pyrene, the major mutagen in tobacco smoke, and DNA

DNA can be damaged by many sorts of mutagens, which change the DNA sequence. Mutagens include oxidizing agents, alkylating agents and also high-energy electromagnetic radiation such as ultraviolet light and X-rays. The type of DNA damage produced depends on the type of mutagen. For example, UV light can damage DNA by producing thymine dimers, which are cross-links between pyrimidine bases. On the other hand, oxidants such as free radicals or hydrogen peroxide produce multiple forms of damage, including base modifications, particularly of guanosine, and double-strand breaks. A typical human cell contains about 150,000 bases that have suffered oxidative damage. Of these oxidative lesions, the most dangerous are double-strand breaks, as these are difficult to repair and can produce point mutations, insertions, deletions from the DNA sequence, and chromosomal translocations. These mutations can cause cancer. Because of inherent limits in the DNA repair mechanisms, if humans lived long enough, they would all eventually develop cancer. DNA damages that are naturally occurring, due to normal cellular processes that produce reactive oxygen species, the hydrolytic activities of cellular water, etc., also occur frequently. Although most of these damages are repaired, in any cell some DNA damage may remain despite the action of repair processes. These remaining DNA damages accumulate with age in mammalian postmitotic tissues. This accumulation appears to be an important underlying cause of aging.

Many mutagens fit into the space between two adjacent base pairs, this is called *intercalation*. Most intercalators are aromatic and planar molecules; examples include ethidium bromide, acridines, daunomycin, and doxorubicin. For an intercalator to fit between base pairs, the bases must separate, distorting the DNA strands by unwinding of the double helix. This inhibits both transcription and DNA replication, causing toxicity and mutations. As a result, DNA intercalators may be carcinogens, and in the case of thalidomide, a teratogen. Others such as benzo[a]pyrene diol epoxide and aflatoxin form DNA adducts that induce errors in replication. Nevertheless, due to their ability to inhibit DNA transcription and replication, other similar toxins are also used in chemotherapy to inhibit rapidly growing cancer cells.

Biological Functions

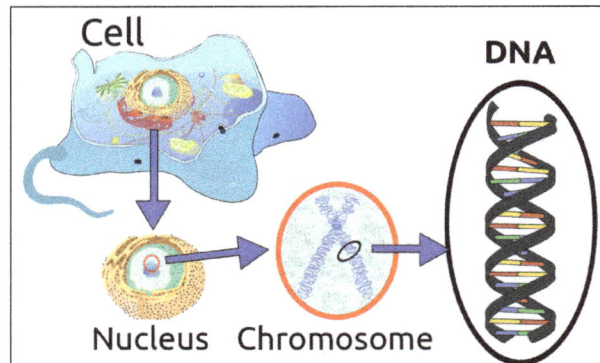

Location of eukaryote nuclear DNA within the chromosomes

DNA usually occurs as linear chromosomes in eukaryotes, and circular chromosomes in prokaryotes. The set of chromosomes in a cell makes up its genome; the human genome has approximately 3 billion base pairs of DNA arranged into 46 chromosomes. The information carried by DNA is held in the sequence of pieces of DNA called genes. Transmission of genetic information in genes is achieved via complementary base pairing. For example, in transcription, when a cell uses the information in a gene, the DNA sequence is copied into a complementary RNA sequence through the attraction between the DNA and the correct RNA nucleotides. Usually, this RNA copy is then used to make a matching protein sequence in a process called translation, which depends on the same interaction between RNA nucleotides. In alternative fashion, a cell may simply copy its genetic information in a process called DNA replication. The details of these functions are covered in other articles; here the focus is on the interactions between DNA and other molecules that mediate the function of the genome.

Genes and Genomes

Genomic DNA is tightly and orderly packed in the process called DNA condensation, to fit the small available volumes of the cell. In eukaryotes, DNA is located in the cell nucleus, with small amounts in mitochondria and chloroplasts. In prokaryotes, the DNA is held within an irregularly shaped body in the cytoplasm called the nucleoid. The genetic information in a genome is held within genes, and the complete set of this information in an organism is called its genotype. A gene is a unit of heredity and is a region of DNA that influences a particular characteristic in an organism. Genes contain an open reading frame that can be transcribed, and regulatory sequences such as promoters and enhancers, which control transcription of the open reading frame.

In many species, only a small fraction of the total sequence of the genome encodes protein. For example, only about 1.5% of the human genome consists of protein-coding exons, with over 50%

of human DNA consisting of non-coding repetitive sequences. The reasons for the presence of so much noncoding DNA in eukaryotic genomes and the extraordinary differences in genome size, or *C-value*, among species, represent a long-standing puzzle known as the "C-value enigma". However, some DNA sequences that do not code protein may still encode functional non-coding RNA molecules, which are involved in the regulation of gene expression.

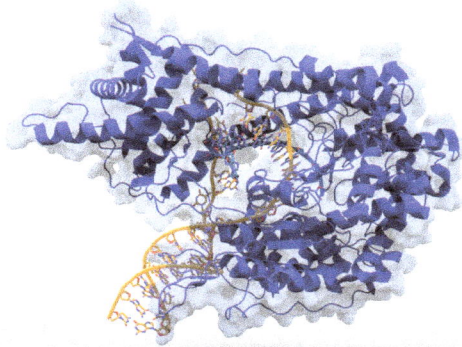

T7 RNA polymerase (blue) producing an mRNA (green) from a DNA template (orange).

Some noncoding DNA sequences play structural roles in chromosomes. Telomeres and centromeres typically contain few genes but are important for the function and stability of chromosomes. An abundant form of noncoding DNA in humans are pseudogenes, which are copies of genes that have been disabled by mutation. These sequences are usually just molecular fossils, although they can occasionally serve as raw genetic material for the creation of new genes through the process of gene duplication and divergence.

Transcription and Translation

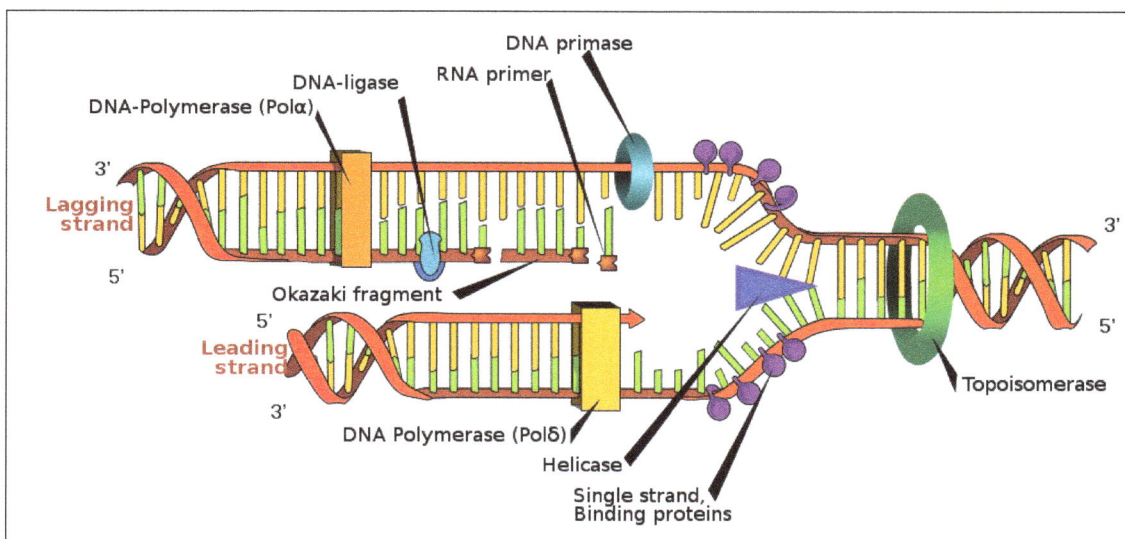

DNA replication: The double helix is unwound by a helicase and topoisomerase. Next, one DNA polymerase produces the leading strand copy. Another DNA polymerase binds to the lagging strand. This enzyme makes discontinuous segments (called Okazaki fragments) before DNA ligase joins them together.

A gene is a sequence of DNA that contains genetic information and can influence the phenotype of an organism. Within a gene, the sequence of bases along a DNA strand defines a messenger RNA sequence, which then defines one or more protein sequences. The relationship between the nucleotide sequences of genes and the amino-acid sequences of proteins is determined by the rules of translation, known collectively as the genetic code. The genetic code consists of three-letter 'words' called *codons* formed from a sequence of three nucleotides (e.g. ACT, CAG, TTT).

In transcription, the codons of a gene are copied into messenger RNA by RNA polymerase. This RNA copy is then decoded by a ribosome that reads the RNA sequence by base-pairing the messenger RNA to transfer RNA, which carries amino acids. Since there are 4 bases in 3-letter combinations, there are 64 possible codons (4^3 combinations). These encode the twenty standard amino acids, giving most amino acids more than one possible codon. There are also three 'stop' or 'nonsense' codons signifying the end of the coding region; these are the TAA, TGA, and TAG codons.

Replication

Cell division is essential for an organism to grow, but, when a cell divides, it must replicate the DNA in its genome so that the two daughter cells have the same genetic information as their parent. The double-stranded structure of DNA provides a simple mechanism for DNA replication. Here, the two strands are separated and then each strand's complementary DNA sequence is recreated by an enzyme called DNA polymerase. This enzyme makes the complementary strand by finding the correct base through complementary base pairing and bonding it onto the original strand. As DNA polymerases can only extend a DNA strand in a 5′ to 3′ direction, different mechanisms are used to copy the antiparallel strands of the double helix. In this way, the base on the old strand dictates which base appears on the new strand, and the cell ends up with a perfect copy of its DNA.

Extracellular Nucleic Acids

Naked extracellular DNA (eDNA), most of it released by cell death, is nearly ubiquitous in the environment. Its concentration in soil may be as high as 2 µg/L, and its concentration in natural aquatic environments may be as high at 88 µg/L. Various possible functions have been proposed for eDNA: it may be involved in horizontal gene transfer; it may provide nutrients; and it may act as a buffer to recruit or titrate ions or antibiotics. Extracellular DNA acts as a functional extracellular matrix component in the biofilms of several bacterial species. It may act as a recognition factor to regulate the attachment and dispersal of specific cell types in the biofilm; it may contribute to biofilm formation; and it may contribute to the biofilm's physical strength and resistance to biological stress.

Cell-free fetal DNA is found in the blood of the mother, and can be sequenced to determine a great deal of information about the developing fetus.

Interactions with Proteins

All the functions of DNA depend on interactions with proteins. These protein interactions can be non-specific, or the protein can bind specifically to a single DNA sequence. Enzymes can also bind to DNA and of these, the polymerases that copy the DNA base sequence in transcription and DNA replication are particularly important.

DNA-binding Proteins

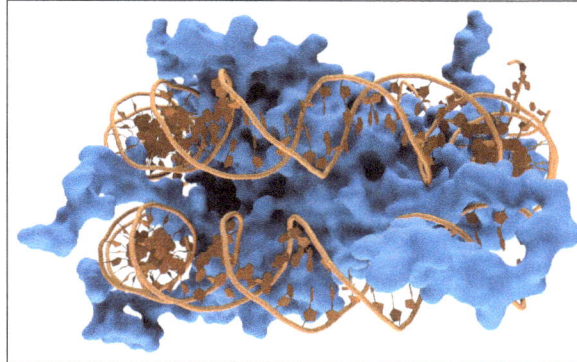

Interaction of DNA (in orange) with histones (in blue). These proteins' basic amino acids bind to the acidic phosphate groups on DNA.

Structural proteins that bind DNA are well-understood examples of non-specific DNA-protein interactions. Within chromosomes, DNA is held in complexes with structural proteins. These proteins organize the DNA into a compact structure called chromatin. In eukaryotes, this structure involves DNA binding to a complex of small basic proteins called histones, while in prokaryotes multiple types of proteins are involved. The histones form a disk-shaped complex called a nucleosome, which contains two complete turns of double-stranded DNA wrapped around its surface. These non-specific interactions are formed through basic residues in the histones, making ionic bonds to the acidic sugar-phosphate backbone of the DNA, and are thus largely independent of the base sequence. Chemical modifications of these basic amino acid residues include methylation, phosphorylation, and acetylation. These chemical changes alter the strength of the interaction between the DNA and the histones, making the DNA more or less accessible to transcription factors and changing the rate of transcription. Other non-specific DNA-binding proteins in chromatin include the high-mobility group proteins, which bind to bent or distorted DNA. These proteins are important in bending arrays of nucleosomes and arranging them into the larger structures that make up chromosomes.

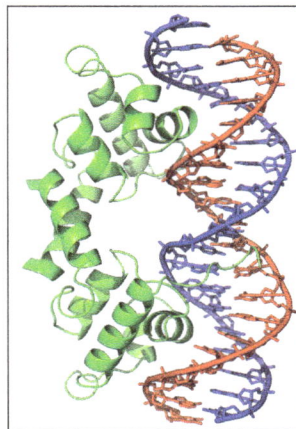

The lambda repressor helix-turn-helix transcription factor bound to its DNA target.

A distinct group of DNA-binding proteins is the DNA-binding proteins that specifically bind single-stranded DNA. In humans, replication protein A is the best-understood member of this family and is used in processes where the double helix is separated, including DNA replication, recombination, and DNA repair. These binding proteins seem to stabilize single-stranded DNA and protect it from forming stem-loops or being degraded by nucleases.

In contrast, other proteins have evolved to bind to particular DNA sequences. The most intensively studied of these are the various transcription factors, which are proteins that regulate transcription. Each transcription factor binds to one particular set of DNA sequences and activates or inhibits the transcription of genes that have these sequences close to their promoters. The transcription factors do this in two ways. Firstly, they can bind the RNA polymerase responsible for transcription, either directly or through other mediator proteins; this locates the polymerase at the promoter and allows it to begin transcription. Alternatively, transcription factors can bind enzymes that modify the histones at the promoter. This changes the accessibility of the DNA template to the polymerase.

As these DNA targets can occur throughout an organism's genome, changes in the activity of one type of transcription factor can affect thousands of genes. Consequently, these proteins are often the targets of the signal transduction processes that control responses to environmental changes or cellular differentiation and development. The specificity of these transcription factors' interactions with DNA come from the proteins making multiple contacts to the edges of the DNA bases, allowing them to "read" the DNA sequence. Most of these base-interactions are made in the major groove, where the bases are most accessible.

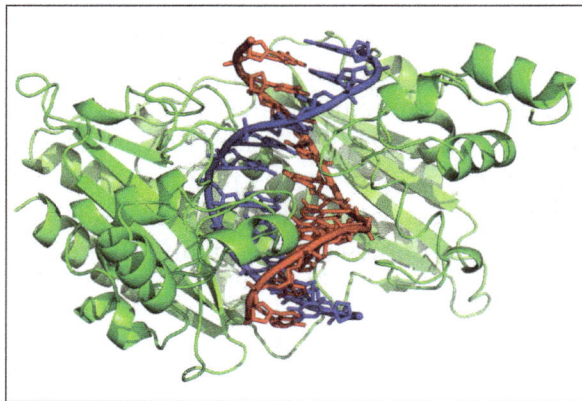

The restriction enzyme EcoRV (green) in a complex with its substrate DNA.

DNA-modifying Enzymes

Nucleases and Ligases

Nucleases are enzymes that cut DNA strands by catalyzing the hydrolysis of the phosphodiester bonds. Nucleases that hydrolyse nucleotides from the ends of DNA strands are called exonucleases, while endonucleases cut within strands. The most frequently used nucleases in molecular biology are the restriction endonucleases, which cut DNA at specific sequences. For instance, the EcoRV enzyme shown to the left recognizes the 6-base sequence 5′-GATATC-3′ and makes a cut at the horizontal line. In nature, these enzymes protect bacteria against phage infection by digesting the phage DNA when it enters the bacterial cell, acting as part of the restriction modification system. In technology, these sequence-specific nucleases are used in molecular cloning and DNA fingerprinting.

Enzymes called DNA ligases can rejoin cut or broken DNA strands. Ligases are particularly important in lagging strand DNA replication, as they join together the short segments of DNA produced at the replication fork into a complete copy of the DNA template. They are also used in DNA repair and genetic recombination.

Topoisomerases and Helicases

Topoisomerases are enzymes with both nuclease and ligase activity. These proteins change the amount of supercoiling in DNA. Some of these enzymes work by cutting the DNA helix and allowing one section to rotate, thereby reducing its level of supercoiling; the enzyme then seals the DNA break. Other types of these enzymes are capable of cutting one DNA helix and then passing a second strand of DNA through this break, before rejoining the helix. Topoisomerases are required for many processes involving DNA, such as DNA replication and transcription.

Helicases are proteins that are a type of molecular motor. They use the chemical energy in nucleoside triphosphates, predominantly adenosine triphosphate (ATP), to break hydrogen bonds between bases and unwind the DNA double helix into single strands. These enzymes are essential for most processes where enzymes need to access the DNA bases.

Polymerases

Polymerases are enzymes that synthesize polynucleotide chains from nucleoside triphosphates. The sequence of their products is created based on existing polynucleotide chains—which are called *templates*. These enzymes function by repeatedly adding a nucleotide to the 3′ hydroxyl group at the end of the growing polynucleotide chain. As a consequence, all polymerases work in a 5′ to 3′ direction. In the active site of these enzymes, the incoming nucleoside triphosphate basepairs to the template: this allows polymerases to accurately synthesize the complementary strand of their template. Polymerases are classified according to the type of template that they use.

In DNA replication, DNA-dependent DNA polymerases make copies of DNA polynucleotide chains. To preserve biological information, it is essential that the sequence of bases in each copy are precisely complementary to the sequence of bases in the template strand. Many DNA polymerases have a proofreading activity. Here, the polymerase recognizes the occasional mistakes in the synthesis reaction by the lack of base pairing between the mismatched nucleotides. If a mismatch is detected, a 3′ to 5′ exonuclease activity is activated and the incorrect base removed. In most organisms, DNA polymerases function in a large complex called the replisome that contains multiple accessory subunits, such as the DNA clamp or helicases.

RNA-dependent DNA polymerases are a specialized class of polymerases that copy the sequence of an RNA strand into DNA. They include reverse transcriptase, which is a viral enzyme involved in the infection of cells by retroviruses, and telomerase, which is required for the replication of telomeres. For example, HIV reverse transcriptase is an enzyme for AIDS virus replication. Telomerase is an unusual polymerase because it contains its own RNA template as part of its structure. It synthesizes telomeres at the ends of chromosomes. Telomeres prevent fusion of the ends of neighboring chromosomes and protect chromosome ends from damage.

Transcription is carried out by a DNA-dependent RNA polymerase that copies the sequence of a DNA strand into RNA. To begin transcribing a gene, the RNA polymerase binds to a sequence of DNA called a promoter and separates the DNA strands. It then copies the gene sequence into a messenger RNA transcript until it reaches a region of DNA called the terminator, where it halts and detaches from the DNA. As with human DNA-dependent DNA polymerases, RNA polymerase II, the enzyme that transcribes most of the genes in the human genome, operates as part of a large protein complex with multiple regulatory and accessory subunits.

Genetic Recombination

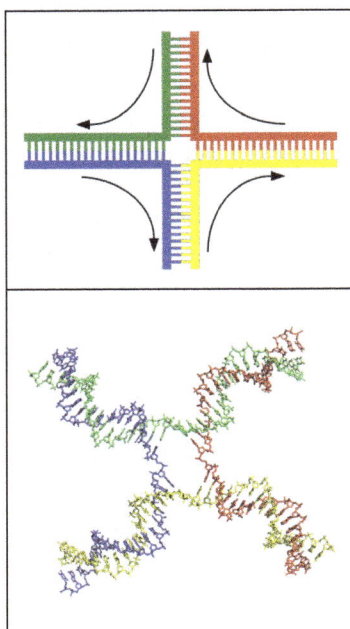

Structure of the Holliday junction intermediate in genetic recombination. The four separate DNA strands are coloured red, blue, green and yellow.

A DNA helix usually does not interact with other segments of DNA, and in human cells, the different chromosomes even occupy separate areas in the nucleus called "chromosome territories". This physical separation of different chromosomes is important for the ability of DNA to function as a stable repository for information, as one of the few times chromosomes interact is in chromosomal crossover which occurs during sexual reproduction, when genetic recombination occurs. Chromosomal crossover is when two DNA helices break, swap a section and then rejoin.

Recombination allows chromosomes to exchange genetic information and produces new combinations of genes, which increases the efficiency of natural selection and can be important in the rapid evolution of new proteins. Genetic recombination can also be involved in DNA repair, particularly in the cell's response to double-strand breaks.

The most common form of chromosomal crossover is homologous recombination, where the two chromosomes involved share very similar sequences. Non-homologous recombination can be damaging to cells, as it can produce chromosomal translocations and genetic abnormalities. The recombination reaction is catalyzed by enzymes known as recombinases, such as RAD51. The first step in recombination is a double-stranded break caused by either an endonuclease or damage to the DNA. A series of steps catalyzed in part by the recombinase then leads to joining of the two helices by at least one Holliday junction, in which a segment of a single strand in each helix is annealed to the complementary strand in the other helix. The Holliday junction is a tetrahedral junction structure that can be moved along the pair of chromosomes, swapping one strand for another. The recombination reaction is then halted by cleavage of the junction and re-ligation of the released DNA. Only strands of like polarity exchange DNA during recombination. There are two types of cleavage: east-west cleavage and north-south cleavage. The north-south cleavage nicks both strands of DNA, while the east-west cleavage has one strand of DNA intact. The formation of a Holliday junction during recombination makes it possible for genetic diversity, genes to exchange on chromosomes, and expression of wild-type viral genomes.

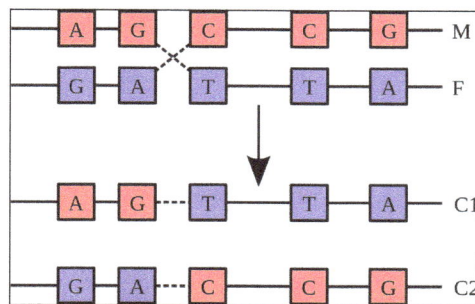

Recombination involves the breaking and rejoining of two chromosomes (M and F) to produce two rearranged chromosomes (C1 and C2).

Evolution

DNA contains the genetic information that allows all forms of life to function, grow and reproduce. However, it is unclear how long in the 4-billion-year history of life DNA has performed this function, as it has been proposed that the earliest forms of life may have used RNA as their genetic material. RNA may have acted as the central part of early cell metabolism as it can both transmit genetic information and carry out catalysis as part of ribozymes. This ancient RNA world where nucleic acid would have been used for both catalysis and genetics may have influenced the evolution of the current genetic code based on four nucleotide bases. This would occur, since the number of different bases in such an organism is a trade-off between a small number of bases increasing replication accuracy and a large number of bases increasing the catalytic efficiency of ribozymes. However, there is no direct evidence of ancient genetic systems, as recovery of DNA from most fossils is impossible because DNA survives in the environment for less than one million years, and slowly degrades into short fragments in solution. Claims for older DNA have been made, most notably a report of the isolation of a viable bacterium from a salt crystal 250 million years old, but these claims are controversial.

Building blocks of DNA (adenine, guanine, and related organic molecules) may have been formed extraterrestrially in outer space. Complex DNA and RNA organic compounds of life, including uracil, cytosine, and thymine, have also been formed in the laboratory under conditions mimicking those found in outer space, using starting chemicals, such as pyrimidine, found in meteorites. Pyrimidine, like polycyclic aromatic hydrocarbons (PAHs), the most carbon-rich chemical found in the universe, may have been formed in red giants or in interstellar cosmic dust and gas clouds.

Uses in Technology

Genetic Engineering

Methods have been developed to purify DNA from organisms, such as phenol-chloroform extraction, and to manipulate it in the laboratory, such as restriction digests and the polymerase chain reaction. Modern biology and biochemistry make intensive use of these techniques in recombinant DNA technology. Recombinant DNA is a man-made DNA sequence that has been assembled from other DNA sequences. They can be transformed into organisms in the form of plasmids or in the appropriate format, by using a viral vector. The genetically modified organisms produced can be used to produce products such as recombinant proteins, used in medical research, or be grown in agriculture.

DNA Profiling

Forensic scientists can use DNA in blood, semen, skin, saliva or hair found at a crime scene to identify a matching DNA of an individual, such as a perpetrator. This process is formally termed DNA profiling, also called *DNA fingerprinting*. In DNA profiling, the lengths of variable sections of repetitive DNA, such as short tandem repeats and minisatellites, are compared between people. This method is usually an extremely reliable technique for identifying a matching DNA. However, identification can be complicated if the scene is contaminated with DNA from several people. DNA profiling was developed in 1984 by British geneticist Sir Alec Jeffreys, and first used in forensic science to convict Colin Pitchfork in the 1988 Enderby murders case.

The development of forensic science and the ability to now obtain genetic matching on minute samples of blood, skin, saliva, or hair has led to re-examining many cases. Evidence can now be uncovered that was scientifically impossible at the time of the original examination. Combined with the removal of the double jeopardy law in some places, this can allow cases to be reopened where prior trials have failed to produce sufficient evidence to convince a jury. People charged with serious crimes may be required to provide a sample of DNA for matching purposes. The most obvious defense to DNA matches obtained forensically is to claim that cross-contamination of evidence has occurred. This has resulted in meticulous strict handling procedures with new cases of serious crime.

DNA profiling is also used successfully to positively identify victims of mass casualty incidents, bodies or body parts in serious accidents, and individual victims in mass war graves, via matching to family members.

DNA profiling is also used in DNA paternity testing to determine if someone is the biological parent or grandparent of a child with the probability of parentage is typically 99.99% when the alleged parent is biologically related to the child. Normal DNA sequencing methods happen after birth, but there are new methods to test paternity while a mother is still pregnant.

DNA Enzymes or Catalytic DNA

Deoxyribozymes, also called DNAzymes or catalytic DNA, were first discovered in 1994. They are mostly single stranded DNA sequences isolated from a large pool of random DNA sequences through a combinatorial approach called in vitro selection or systematic evolution of ligands by exponential enrichment (SELEX). DNAzymes catalyze variety of chemical reactions including

RNA-DNA cleavage, RNA-DNA ligation, amino acids phosphorylation-dephosphorylation, carbon-carbon bond formation, and etc. DNAzymes can enhance catalytic rate of chemical reactions up to 100,000,000,000-fold over the uncatalyzed reaction. The most extensively studied class of DNAzymes is RNA-cleaving types which have been used to detect different metal ions and designing therapeutic agents. Several metal-specific DNAzymes have been reported including the GR-5 DNAzyme (lead-specific), the CA1-3 DNAzymes (copper-specific), the 39E DNAzyme (uranyl-specific) and the NaA43 DNAzyme (sodium-specific). The NaA43 DNAzyme, which is reported to be more than 10,000-fold selective for sodium over other metal ions, was used to make a real-time sodium sensor in cells.

Bioinformatics

Bioinformatics involves the development of techniques to store, data mine, search and manipulate biological data, including DNA nucleic acid sequence data. These have led to widely applied advances in computer science, especially string searching algorithms, machine learning, and database theory. String searching or matching algorithms, which find an occurrence of a sequence of letters inside a larger sequence of letters, were developed to search for specific sequences of nucleotides. The DNA sequence may be aligned with other DNA sequences to identify homologous sequences and locate the specific mutations that make them distinct. These techniques, especially multiple sequence alignment, are used in studying phylogenetic relationships and protein function. Data sets representing entire genomes' worth of DNA sequences, such as those produced by the Human Genome Project, are difficult to use without the annotations that identify the locations of genes and regulatory elements on each chromosome. Regions of DNA sequence that have the characteristic patterns associated with protein- or RNA-coding genes can be identified by gene finding algorithms, which allow researchers to predict the presence of particular gene products and their possible functions in an organism even before they have been isolated experimentally. Entire genomes may also be compared, which can shed light on the evolutionary history of particular organism and permit the examination of complex evolutionary events.

DNA Nanotechnology

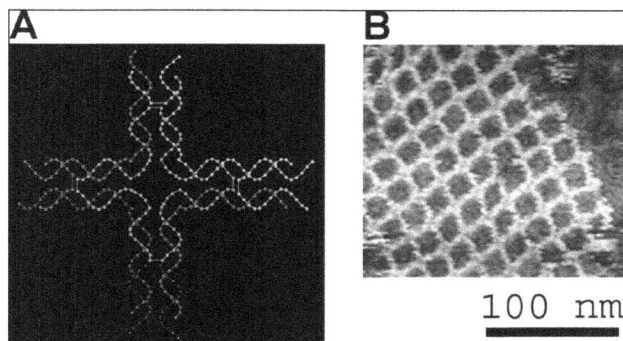

The DNA structure at left (schematic shown) will self-assemble into the structure visualized by atomic force microscopy at right. DNA nanotechnology is the field that seeks to design nanoscale structures using the molecular recognition properties of DNA molecules.

DNA nanotechnology uses the unique molecular recognition properties of DNA and other nucleic acids to create self-assembling branched DNA complexes with useful properties. DNA is thus used as a structural material rather than as a carrier of biological information. This has led to the creation of two-dimensional periodic lattices (both tile-based and using the *DNA origami* method) and three-dimensional structures in the shapes of polyhedra. Nanomechanical devices and algorithmic self-assembly have also been demonstrated, and these DNA structures have been used to template the arrangement of other molecules such as gold nanoparticles and streptavidin proteins.

History and Anthropology

Because DNA collects mutations over time, which are then inherited, it contains historical information, and, by comparing DNA sequences, geneticists can infer the evolutionary history of organisms, their phylogeny. This field of phylogenetics is a powerful tool in evolutionary biology. If DNA sequences within a species are compared, population geneticists can learn the history of particular populations. This can be used in studies ranging from ecological genetics to anthropology.

Information Storage

DNA as a storage device for information has enormous potential since it has much higher storage density compared to electronic devices. However high costs, extremely slow read and write times (memory latency), and insufficient reliability has prevented its practical use.

Plasmid

Plasmids are small, circular molecules of DNA that are capable of replicating independently. As such, they do not rely on chromosomal DNA of the organism for replication. Because of this characteristic, they are also referred to as extra-chromosomal DNA.

Although the molecule was first discovered in a member of the Enterobacteriacae, studies have shown that plasmids are naturally occurring in many types of microorganisms around the world.

Although there have been debates as to whether plasmids can be regarded as microorganisms, at least using the proposed virus definition, it is worth noting that the term "plasmid" is largely used to refer to genetic elements that exist outside the chromosome (in DNA of the organism) and are capable of replicating independently.

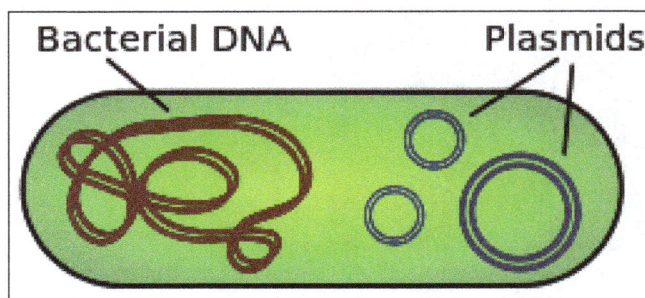

Bacterium with its chromosomal DNA and several plasmids within it.

Structure

With regards to structure, plasmids are made up of circular double chains of DNA. The circular structure of plasmids is made possible by the two ends of the double strands being joined by covalent bonds. The molecules are also small in size, especially when compared to the organisms' DNA, and measure between a few kilobases and several hundred kilobases.

Although a good number of plasmids have a covalently closed circular structure, some plasmids have a linear structure and do not form a circular shape Generally, plasmids are composed of three major components that include:

1. Origin of replication (replicon) - The origin of replication (ori) refers to a specific location in the strand at which replication begins. For plasmids, this location is largely composed of A-T base pairs that are easier to separate during replication.

Compared to the organisms' DNA that consists of many origins of replication, plasmids have one of a few origins of replication because they are smaller in size. At the origin of replication, plasmids also contain a number of regulatory elements that contribute to the process (e.g. Rep proteins)

2. Polylinker (multiple cloning sites) - In plasmid, the polylinker (MCS) is one of the most important parts of the molecule. This is because it allows students to learn more about cloning. Basically, a polylinker is a short sequence of DNA consisting of a few sites for cleavage by restriction enzymes.

As such, MCS allows for easy insertion of DNA through ligation or restriction enzyme digestion. At the site of cleavage, different polylinkers can cut the strand. Therefore, one of the restriction enzymes can cut the plasmid at given points of the sire to allow for DNA insertion.

3. Antibiotic resistance gene - The antibiotic resistance gene is one of the main components of plasmids. These genes play an important role in drug resistance (to one or more antibiotics) thus making treatment of some diseases more challenging.

Plasmids are today known for their ability to transfer from one species of bacteria to another through a process known as conjugation (contact between cells that is followed by transfer of DNA content). In the process, they are capable of conferring antibiotic resistance properties to other species of bacteria.

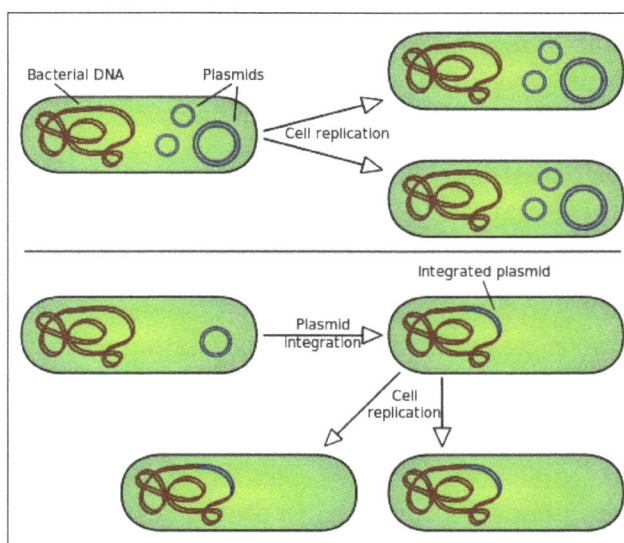

Comparing the activity of non-integrating plasmids, on the top, with episomes, on the bottom, during cell division by Spaully.

While plasmid replication provides an added advantage to bacteria (resistance to certain antibiotics), it also affects cell division of bacteria due to additional replication burden. As a result, bacteria with plasmids tend to be out populated by those without plasmids due to reduced cell division.

Some of the other components of plasmids include:

1. A promoter region - This is the component of plasmids that is involved in recruiting transcriptional machinery.

2. Primer binding site - This is a short sequence of DNA on a single strand that is typically used for the purposes of PCR amplification or DNA sequencing.

Although plasmids share various general characteristics, there are different types in existence.

Types and Functions of Plasmids

Resistance Plasmids

Also referred to as antimicrobial resistance plasmids, resistance plasmids are a type of plasmids that carry genes that play an important role in antibiotic resistance. They are also highly involved in bacterial conjugation by producing conjugation pili which transfer the R plasmid from one bacterium to another.

Resistance Plasmids are divided into two main groups that include:

1. Narrow-host-range group - Often replicated within a single species.

2. Broad-host-range group - Easily transferred between bacteria species. This group of resistance plasmids has been shown to carry a range of antibiotic resistance genes. Following the transfer of antibiotic resistance genes to drug-sensitive bacteria, this can cause the bacteria to develop resistance towards a variety of drugs.

Compared to other types of plasmids, degradative plasmids enable the host organism to degrade/break down xenobiotic compounds. Also referred to as recalcitrant substances, xenobiotic compounds include a range of compounds released into the environment as a result of human actions and are therefore not naturally occurring or common in nature.

Hosts of degradative plasmids are found in groups IncP-1, IncP-7, and IncP-9 and include such species as Ochrobactrum anthropi, Rhizobium sp, Burkholderia hospita, Escherichia coli, and Pseudomonas fluorescens among many others.

Because of the ability of the host to degrade xenobiotic compounds, researchers have attempted to use the plasmids to degrade various contaminating substances in the environment. However, given that this has not proved effective, research studies continue to be conducted to determine how to use various indigenous bacteria (as hosts of degradative plasmids) for degradation of such compounds.

While degradative plasmids contribute to the degradation of xenobiotic compounds, their behavior varies depending on a number of factors such as the capacity for replication and stability. For instance, plasmids found in IncP-1 group have not only been shown to have a broad host range, but also high transfer frequency.

The differences in the behavior of different degradative plasmids have therefore been shown to result in different behaviors between them and their respective hosts.

The use of biodegradative microorganisms for the purposes of removing xenobiotic compounds from contaminated environments is known as Bioaugmentation.

Whereas IncP-1 plasmids have a broad range of hosts, IncP-7 has been shown to have a narrow host range. On the other hand, IncP-9 has an intermediate host range.

Fertility Plasmids

Like many other plasmids, fertility plasmids (F plasmid) have a circular structure and measures about 100 kb.

Some of the main parts of the F plasmid include:

- Transposable element (IS2, 1S3, and Tn1000),

- Replication sites (RepFIA, RepFIB, and RepFIC),

- Origin of conjugative transfer (oirT),

- Replication origin regions.

F plasmid plays an important role in reproduction given that they contain genes that code for the production of sex pilus as well as enzymes required for conjugation. F plasmid also contains genes that are involved in their own transfer. Therefore, during conjugation, they enhance their own transfer from one cell to another.

Whereas the cells that process the F plasmids are referred to as donors, those that lack this factor are the recipients. On the other hand, the plasmids that enhance the ability of the host cell to be-have like a donor are known as the transfer factor.

During conjugation, the donor cell (bacteria) with sex pili (1-3 sex pili) binds to a specific protein on the outer membrane of the recipient thus initiating the mating process.

Following the initial binding, the pili retract thus allowing the two cells to bind together. This is then followed by the transfer of DNA from the donor to the recipient and consequently the transfer of the F plasmid. As a result, the recipient acquires the F factor and gains the ability to produce sex pilus involved in conjugation.

During conjugation, only DNA is passed from the donor to the recipient. Therefore, cytoplasm and other cell material are not transferred.

Sexual pili (sex pili) are tiny rod-like structures that allow the F-positive (cells that have the F fac-tor) bacterial cells to attach to the F-negative (cells lacking the pili) female to promote conjugative transfer.

Col Plasmids

Col plasmids confer to bacteria the ability to produce toxic proteins known as colicines. Such bacteria as E. coli, Shigella and Salmonella use these toxins to kill other bacteria and thus thrive in their respective environments.

There are different types of Col plasmids in existence that produce different types of colicines/colicins. A few examples of Col plasmids include Col B, Col E2 and E3. Their differences are also characterized by differences in their mode of action.

For instance, whereas Col B causes damage to cell membrane of other bacteria (lacking the plasmid) Col E3 has been shown to induce degradation of the nucleic acids of the target cells.

Like fertility plasmids, some of the Col plasmids have been shown to carry elements that enhance their transmission from one cell to another. Therefore, through conjugation or the mating process, particularly for cells with the F factor (fertility plasmids) the Col plasmids can be transferred from one cell (donor) to another (recipient). As a result, the recipient acquires the ability to produce toxins that kill or inhibit the growth of the target bacteria lacking the plasmid.

Colicins/colicines belong to a group of toxins known as bacteriocins.

These toxins affect the target bacteria by affecting such processes as replication of DNA, translation and energy metabolism among others.

Virulence Plasmids

Compared to other harmless bacteria, bacteria that tend to be pathogenic in nature carry genes for virulence factors that allow them to invade and infect their respective hosts.

For some of these bacteria, the virulence factors are the result of the organisms' own genetic material. However, for others, this is as a result of genetic elements from extra-chromosomal DNA. Although there are other sources of such elements, e.g. transposons, plasmids are some of the most common mobile genetic elements.

With regards to pathogenicity, virulence plasmids play an important role given that they can help bacteria effectively adapt to their respective environments. This is because the virulence plasmid can enable the organism to express an array of virulence-associated functions thus providing the organism with more advantageous characteristics to thrive in their environment.

Like other types of plasmids, virulence plasmids can also be transmitted from one bacterium to another. Apart from the virulence gene, plasmids have also been shown to carry other important elements that enhance transmission and maintenance.

For this reason, they are larger in size but low in numbers. This ensures that they do not cause additional burden to the organism during cell division.

Typically, cell division and cell maintenance require the use of energy. By having low numbers of virulence plasmids, the cells are spared significant metabolic burden that would be required for maintenance and genome duplication of numerous plasmids.

Some of the other types of plasmids include:

- Recombinant plasmids - Plasmids that have been altered in the laboratory and introduced into the bacteria for the purposes of studies.

- Crptic plasmids - No known functions.

- Metabolic plasmids - Enhance metabolism of the host.

- Conjugative plasmids - Promote self-transfer.

- Suicide plasmids - Fail to replicate when transferred from one cell to another.

Plasmid Vector

A vector refers to any piece of molecule that contains genetic material that can be replicated and expressed when transferred into another cell. Based on this definition, it is possible to see why the words "vector" and "plasmids" are sometimes interchanged. However, this is not to say that all plasmids are vectors.

One of the primary characteristics of plasmid vectors is that they are small in size. Apart from their size, they are characterized by an origin of replication, a selective marker as well as multiple cloning sites.

The ideal plasmid vectors have high copy numbers inside the cell. As such, it ensures high numbers of the target gene for cloning purposes. This also ensures that the gene of interest is increased during genomic division. In addition, the plasmid can have a marker gene as the visual marker to help determine whether cloning was successful.

Because of their multiple cloning sites, plasmids have been shown to be some of the best vectors for cloning. Because of this characteristic, it is possible for restriction enzymes to cleave various regions of the plasmid for cloning.

Over the years, using these vectors has allowed for recombinant DNA to be introduced into host cells for study purposes. For instance, through this type of cloning, it has become possible for researchers to sequence the genome of a range of species, study the expression of genes and even observe various cellular mechanisms.

While smaller plasmids are capable of carrying long DNA segments, reduced size can also help remove non-essential genes that are not required for cloning.

Plasmid Isolation

In order to obtain purified plasmid DNA for such procedures as cloning, PCR and transfection, plasmid isolation has to be performed. The process involves using a number of techniques to obtain the plasmid DNA from host cells in order to use it in molecular biology.

Plasmid isolation involves the following steps:

- Cell growth (growth of bacterial cells) - This involves growing the bacteria that contain plasmid in a specific shaken culture. Here, given antibiotics may be used to prevent the growth of other undesired bacteria.

- Centrifugation - Bacterial growth is followed by centrifugation in order to pellet the cells. Once the supernatant has been removed, then isolation of plasmids can begin.

One of the most common techniques for isolation is the classical method that is sometimes referred to as alkaline lysis.

This involves the following steps:

- Suspension of the pellet in an isotonic solution - The bacterial pellet obtained from centrifugation is re-suspended in an isotonic solution (ethylene diamine tetraacetate) which prevents nuclease activity.

- Alkaline lysis of the cells - This involves cell lysis by using sodium dodecyl sulfate to disintegrate the lipid structure on the cell membrane.

- Precipitation of dissolved proteins using a solution of acidic potassium acetate.

- Sedimentation - centrifugation is used for sedimentation.

- Purification - A mixture of phenol and chloroform is used for purification of the plasmid DNA. This step removes protein content.

- Add ethanol for precipitation (to be sedimented through centrifugation).

- Wash the solution with 70 percent ethanol (to remove salt content).

- Centrifuge to sediment plasmid DNA.

- Dissolve in TE solution and store.

RNA

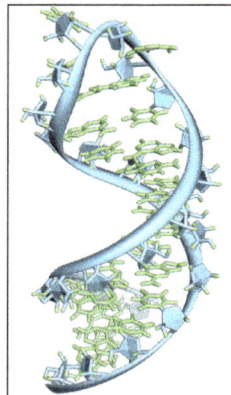

A hairpin loop from a pre-mRNA. Highlighted are the nucleobases (green) and the ribose-phosphate backbone (blue). This is a single strand of RNA that folds back upon itself.

Ribonucleic acid (RNA) is a polymeric molecule essential in various biological roles in coding, decoding, regulation and expression of genes. RNA and DNA are nucleic acids, and, along with lipids, proteins and carbohydrates, constitute the four major macromolecules essential for all known forms of life. Like DNA, RNA is assembled as a chain of nucleotides, but unlike DNA it is more often found in nature as a single-strand folded onto itself, rather than a paired double-strand.

Cellular organisms use messenger RNA (*mRNA*) to convey genetic information (using the nitrogenous bases of guanine, uracil, adenine, and cytosine, denoted by the letters G, U, A, and C) that directs synthesis of specific proteins. Many viruses encode their genetic information using an RNA genome.

Some RNA molecules play an active role within cells by catalyzing biological reactions, controlling gene expression, or sensing and communicating responses to cellular signals. One of these active processes is protein synthesis, a universal function in which RNA molecules direct the synthesis of proteins on ribosomes. This process uses transfer RNA (*tRNA*) molecules to deliver amino acids to the ribosome, where ribosomal RNA (*rRNA*) then links amino acids together to form coded proteins.

Comparison with DNA

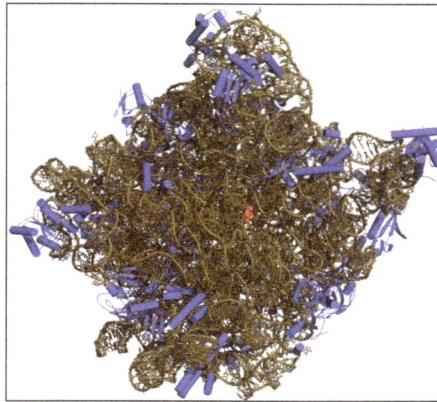

Three-dimensional representation of the 50S ribosomal subunit. Ribosomal RNA is in ochre, proteins in blue. The active site is a small segment of rRNA, indicated in red.

The chemical structure of RNA is very similar to that of DNA, but differs in three primary ways:

- Unlike double-stranded DNA, RNA is a single-stranded molecule in many of its biological roles and consists of much shorter chains of nucleotides. However, a single RNA molecule can, by complementary base pairing, form intrastrand double helixes, as in tRNA.

- While the sugar-phosphate "backbone" of DNA contains *deoxyribose*, RNA contains *ribose* instead. Ribose has a hydroxyl group attached to the pentose ring in the 2' position, whereas deoxyribose does not. The hydroxyl groups in the ribose backbone make RNA more chemically labile than DNA by lowering the activation energy of hydrolysis.

- The complementary base to adenine in DNA is thymine, whereas in RNA, it is uracil, which is an unmethylated form of thymine.

Like DNA, most biologically active RNAs, including mRNA, tRNA, rRNA, snRNAs, and other non-coding RNAs, contain self-complementary sequences that allow parts of the RNA to fold and pair with itself to form double helices. Analysis of these RNAs has revealed that they are highly structured. Unlike DNA, their structures do not consist of long double helices, but rather collections of short helices packed together into structures akin to proteins. In this fashion, RNAs can achieve chemical catalysis (like enzymes). For instance, determination of the structure of the ribosome—an RNA-protein complex that catalyzes peptide bond formation—revealed that its active site is composed entirely of RNA.

Structure

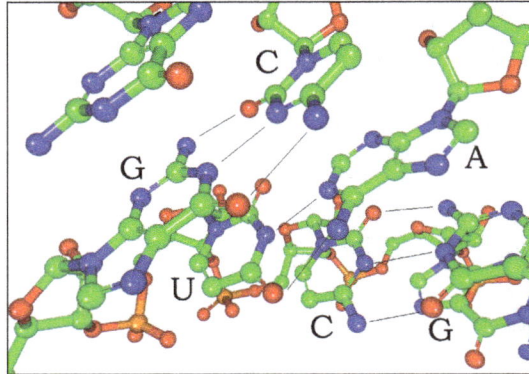

Watson-Crick base pairs in a siRNA (hydrogen atoms are not shown).

Each nucleotide in RNA contains a ribose sugar, with carbons numbered 1' through 5'. A base is attached to the 1' position, in general, adenine (A), cytosine (C), guanine (G), or uracil (U). Adenine and guanine are purines, cytosine and uracil are pyrimidines. A phosphate group is attached to the 3' position of one ribose and the 5' position of the next. The phosphate groups have a negative charge each, making RNA a charged molecule (polyanion). The bases form hydrogen bonds between cytosine and guanine, between adenine and uracil and between guanine and uracil. However, other interactions are possible, such as a group of adenine bases binding to each other in a bulge, or the GNRA tetraloop that has a guanine–adenine base-pair.

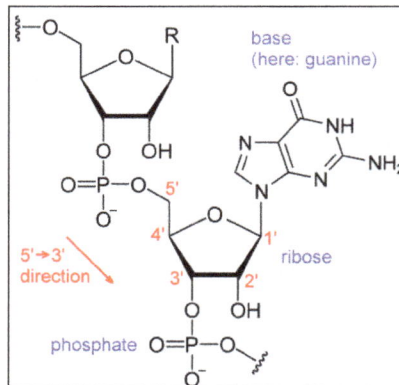

Chemical structure of RNA.

An important structural component of RNA that distinguishes it from DNA is the presence of a hydroxyl group at the 2' position of the ribose sugar. The presence of this functional group causes the helix to mostly take the A-form geometry, although in single strand dinucleotide contexts, RNA can rarely also adopt the B-form most commonly observed in DNA. The A-form geometry results in a very deep and narrow major groove and a shallow and wide minor groove. A second consequence of the presence of the 2'-hydroxyl group is that in conformationally flexible regions of an RNA molecule (that is, not involved in formation of a double helix), it can chemically attack the adjacent phosphodiester bond to cleave the backbone.

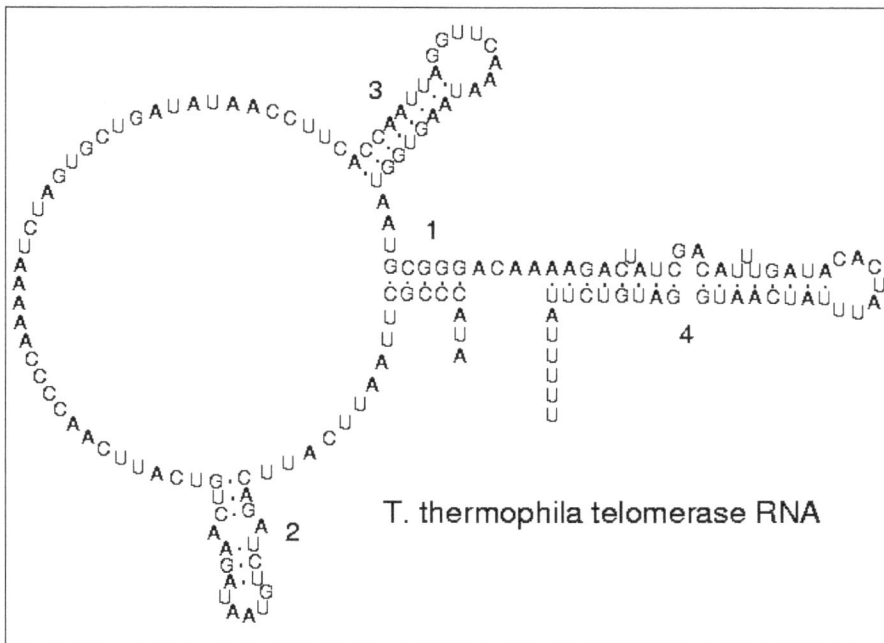

Secondary structure of a telomerase RNA.

RNA is transcribed with only four bases (adenine, cytosine, guanine and uracil), but these bases and attached sugars can be modified in numerous ways as the RNAs mature. Pseudouridine (Ψ), in which the linkage between uracil and ribose is changed from a C–N bond to a C–C bond, and ribothymidine (T) are found in various places (the most notable ones being in the TΨC loop of tRNA). Another notable modified base is hypoxanthine, a deaminated adenine base whose nucleoside is called inosine (I). Inosine plays a key role in the wobble hypothesis of the genetic code.

There are more than 100 other naturally occurring modified nucleosides. The greatest structural diversity of modifications can be found in tRNA, while pseudouridine and nucleosides with 2'-O-methylribose often present in rRNA are the most common. The specific roles of many of these modifications in RNA are not fully understood. However, it is notable that, in ribosomal RNA, many of the post-transcriptional modifications occur in highly functional regions, such as the peptidyl transferase center and the subunit interface, implying that they are important for normal function.

The functional form of single-stranded RNA molecules, just like proteins, frequently requires a specific tertiary structure. The scaffold for this structure is provided by secondary structural elements that are hydrogen bonds within the molecule. This leads to several recognizable "domains" of secondary structure like hairpin loops, bulges, and internal loops. Since RNA is charged, metal ions such as Mg^{2+} are needed to stabilise many secondary and tertiary structures.

The naturally occurring enantiomer of RNA is D-RNA composed of D-ribonucleotides. All chirality centers are located in the D-ribose. By the use of L-ribose or rather L-ribonucleotides, L-RNA can be synthesized. L-RNA is much more stable against degradation by RNase.

Like other structured biopolymers such as proteins, one can define topology of a folded RNA molecule. This is often done based on arrangement of intra-chain contacts within a folded RNA, termed as circuit topology.

Synthesis

Synthesis of RNA is usually catalyzed by an enzyme—RNA polymerase—using DNA as a template, a process known as transcription. Initiation of transcription begins with the binding of the enzyme to a promoter sequence in the DNA (usually found "upstream" of a gene). The DNA double helix is unwound by the helicase activity of the enzyme. The enzyme then progresses along the template strand in the 3' to 5' direction, synthesizing a complementary RNA molecule with elongation occurring in the 5' to 3' direction. The DNA sequence also dictates where termination of RNA synthesis will occur.

Primary transcript RNAs are often modified by enzymes after transcription. For example, a poly(A) tail and a 5' cap are added to eukaryotic pre-mRNA and introns are removed by the spliceosome.

There are also a number of RNA-dependent RNA polymerases that use RNA as their template for synthesis of a new strand of RNA. For instance, a number of RNA viruses (such as poliovirus) use this type of enzyme to replicate their genetic material. Also, RNA-dependent RNA polymerase is part of the RNA interference pathway in many organisms.

Types of RNA

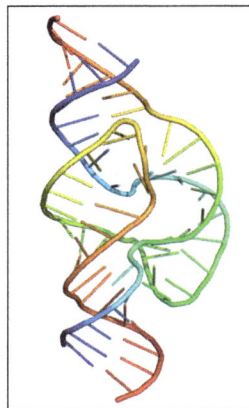

Structure of a hammerhead ribozyme, a ribozyme that cuts RNA.

Messenger RNA (mRNA) is the RNA that carries information from DNA to the ribosome, the sites of protein synthesis (translation) in the cell. The coding sequence of the mRNA determines the amino acid sequence in the protein that is produced. However, many RNAs do not code for protein (about 97% of the transcriptional output is non-protein-coding in eukaryotes).

These so-called non-coding RNAs ("ncRNA") can be encoded by their own genes (RNA genes), but can also derive from mRNA introns. The most prominent examples of non-coding RNAs are transfer RNA (tRNA) and ribosomal RNA (rRNA), both of which are involved in the process of translation. There are also non-coding RNAs involved in gene regulation, RNA processing and other roles. Certain RNAs are able to catalyse chemical reactions such as cutting and ligating other RNA molecules, and the catalysis of peptide bond formation in the ribosome; these are known as ribozymes.

In length

According to the length of RNA chain, RNA includes small RNA and long RNA. Usually, small

RNAs are shorter than 200 nt in length, and long RNAs are greater than 200 nt long. Long RNAs, also called large RNAs, mainly include long non-coding RNA (lncRNA) and mRNA. Small RNAs mainly include 5.8S ribosomal RNA (rRNA), 5S rRNA, transfer RNA (tRNA), microRNA (miRNA), small interfering RNA (siRNA), small nucleolar RNA (snoRNAs), Piwi-interacting RNA (piRNA), tRNA-derived small RNA (tsRNA) and small rDNA-derived RNA (srRNA).

In Translation

Messenger RNA (mRNA) carries information about a protein sequence to the ribosomes, the protein synthesis factories in the cell. It is coded so that every three nucleotides (a codon) corresponds to one amino acid. In eukaryotic cells, once precursor mRNA (pre-mRNA) has been transcribed from DNA, it is processed to mature mRNA. This removes its introns—non-coding sections of the pre-mRNA. The mRNA is then exported from the nucleus to the cytoplasm, where it is bound to ribosomes and translated into its corresponding protein form with the help of tRNA. In prokaryotic cells, which do not have nucleus and cytoplasm compartments, mRNA can bind to ribosomes while it is being transcribed from DNA. After a certain amount of time, the message degrades into its component nucleotides with the assistance of ribonucleases.

Transfer RNA (tRNA) is a small RNA chain of about 80 nucleotides that transfers a specific amino acid to a growing polypeptide chain at the ribosomal site of protein synthesis during translation. It has sites for amino acid attachment and an anticodon region for codon recognition that binds to a specific sequence on the messenger RNA chain through hydrogen bonding.

Ribosomal RNA (rRNA) is the catalytic component of the ribosomes. Eukaryotic ribosomes contain four different rRNA molecules: 18S, 5.8S, 28S and 5S rRNA. Three of the rRNA molecules are synthesized in the nucleolus, and one is synthesized elsewhere. In the cytoplasm, ribosomal RNA and protein combine to form a nucleoprotein called a ribosome. The ribosome binds mRNA and carries out protein synthesis. Several ribosomes may be attached to a single mRNA at any time. Nearly all the RNA found in a typical eukaryotic cell is rRNA.

Transfer-messenger RNA (tmRNA) is found in many bacteria and plastids. It tags proteins encoded by mRNAs that lack stop codons for degradation and prevents the ribosome from stalling.

Regulatory RNA

The earliest known regulators of gene expression were proteins known as repressors and activators, regulators with specific short binding sites within enhancer regions near the genes to be regulated. More recently, RNAs have been found to regulate genes as well. There are several kinds of RNA-dependent processes in eukaryotes regulating the expression of genes at various points, such as RNAi repressing genes post-transcriptionally, long non-coding RNAs shutting down blocks of chromatin epigenetically, and enhancer RNAs inducing increased gene expression. In addition to these mechanisms in eukaryotes, both bacteria and archaea have been found to use regulatory RNAs extensively. Bacterial small RNA and the CRISPR system are examples of such prokaryotic regulatory RNA systems. Fire and Mello were awarded the 2006 Nobel Prize in Physiology or Medicine for discovering microRNAs (miRNAs), specific short RNA molecules that can base-pair with mRNAs.

RNA Interference by miRNAs

Post-transcriptional expression levels of many genes can be controlled by RNA interference, in which miRNAs, specific short RNA molecules, pair with meRNA regions and target them for degradation. This antisense-based process involves steps that first process the RNA so that it can base-pair with a region of its target mRNAs. Once the base pairing occurs, other proteins direct the mRNA to be destroyed by nucleases. Fire and Mello were awarded the 2006 Nobel Prize in Physiology or Medicine for this discovery.

Long Non-coding RNAs

Next to be linked to regulation were Xist and other long noncoding RNAs associated with X chromosome inactivation. Their roles, at first mysterious, were shown by Jeannie T. Lee and others to be the silencing of blocks of chromatin via recruitment of Polycomb complex so that messenger RNA could not be transcribed from them. Additional lncRNAs, currently defined as RNAs of more than 200 base pairs that do not appear to have coding potential, have been found associated with regulation of stem cell pluripotency and cell division.

Enhancer RNAs

The third major group of regulatory RNAs is called enhancer RNAs. It is not clear at present whether they are a unique category of RNAs of various lengths or constitute a distinct subset of lncRNAs. In any case, they are transcribed from enhancers, which are known regulatory sites in the DNA near genes they regulate. They up-regulate the transcription of the gene(s) under control of the enhancer from which they are transcribed.

Regulatory RNA in Prokaryotes

At first, regulatory RNA was thought to be a eukaryotic phenomenon, a part of the explanation for why so much more transcription in higher organisms was seen than had been predicted. But as soon as researchers began to look for possible RNA regulators in bacteria, they turned up there as well. Currently, the ubiquitous nature of systems of RNA regulation of genes has been discussed as support for the RNA World theory. Bacterial small RNAs generally act via antisense pairing with mRNA to down-regulate its translation, either by affecting stability or affecting cis-binding ability. Riboswitches have also been discovered. They are cis-acting regulatory RNA sequences acting allosterically. They change shape when they bind metabolites so that they gain or lose the ability to bind chromatin to regulate expression of genes.

Archaea also have systems of regulatory RNA. The CRISPR system, recently being used to edit DNA *in situ*, acts via regulatory RNAs in archaea and bacteria to provide protection against virus invaders.

In RNA Processing

Many RNAs are involved in modifying other RNAs. Introns are spliced out of pre-mRNA by spliceosomes, which contain several small nuclear RNAs (snRNA), or the introns can be ribozymes that are spliced by themselves. RNA can also be altered by having its nucleotides modified to nucleotides other than A, C, G and U. In eukaryotes, modifications of RNA nucleotides are in general directed by small nucleolar RNAs (snoRNA; 60–300 nt), found in the nucleolus and cajal bodies. snoRNAs associate with enzymes and guide them to a spot on an RNA by basepairing to that RNA. These enzymes then perform the nucleotide modification. rRNAs and tRNAs are extensively modified, but snRNAs and mRNAs can also be the target of base modification. RNA can also be methylated.

Uridine to pseudouridine is a common RNA modification.

RNA Genomes

Like DNA, RNA can carry genetic information. RNA viruses have genomes composed of RNA that encodes a number of proteins. The viral genome is replicated by some of those proteins, while other proteins protect the genome as the virus particle moves to a new host cell. Viroids are another group of pathogens, but they consist only of RNA, do not encode any protein and are replicated by a host plant cell's polymerase.

In Reverse Transcription

Reverse transcribing viruses replicate their genomes by reverse transcribing DNA copies from their RNA; these DNA copies are then transcribed to new RNA. Retrotransposons also spread by copying DNA and RNA from one another, and telomerase contains an RNA that is used as template for building the ends of eukaryotic chromosomes.

Double-stranded RNA

Double-stranded RNA.

Double-stranded RNA (dsRNA) is RNA with two complementary strands, similar to the DNA found in all cells, but with the replacement of thymine by uracil. dsRNA forms the genetic material of some viruses (double-stranded RNA viruses). Double-stranded RNA, such as viral RNA or siRNA, can trigger RNA interference in eukaryotes, as well as interferon response in vertebrates.

Circular RNA

In the late 1970s, it was shown that there is a single stranded covalently closed, i.e. circular form of RNA expressed throughout the animal and plant kingdom circRNAs are thought to arise via a "back-splice" reaction where the spliceosome joins a downstream donor to an upstream acceptor splice site. So far the function of circRNAs is largely unknown, although for few examples a microRNA sponging activity has been demonstrated.

Key Discoveries in RNA Biology

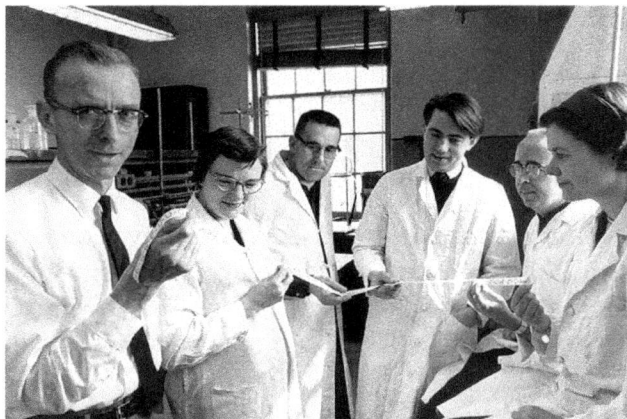

Robert W. Holley, left, poses with his research team.

Research on RNA has led to many important biological discoveries and numerous Nobel Prizes. Nucleic acids were discovered in 1868 by Friedrich Miescher, who called the material 'nuclein' since it was found in the nucleus. It was later discovered that prokaryotic cells, which do not have a nucleus, also contain nucleic acids. The role of RNA in protein synthesis was suspected already in 1939. Severo Ochoa won the 1959 Nobel Prize in Medicine (shared with Arthur Kornberg) after he discovered an enzyme that can synthesize RNA in the laboratory. However, the enzyme discovered by Ochoa (polynucleotide phosphorylase) was later shown to be responsible for RNA degradation, not RNA synthesis. In 1956 Alex Rich and David Davies hybridized two separate strands of RNA to form the first crystal of RNA whose structure could be determined by X-ray crystallography.

The sequence of the 77 nucleotides of a yeast tRNA was found by Robert W. Holley in 1965, winning Holley the 1968 Nobel Prize in Medicine (shared with Har Gobind Khorana and Marshall Nirenberg).

During the early 1970s, retroviruses and reverse transcriptase were discovered, showing for the first time that enzymes could copy RNA into DNA (the opposite of the usual route for transmission of genetic information). For this work, David Baltimore, Renato Dulbecco and

Howard Temin were awarded a Nobel Prize in 1975. In 1976, Walter Fiers and his team determined the first complete nucleotide sequence of an RNA virus genome, that of bacteriophage MS2.

In 1977, introns and RNA splicing were discovered in both mammalian viruses and in cellular genes, resulting in a 1993 Nobel to Philip Sharp and Richard Roberts. Catalytic RNA molecules (ribozymes) were discovered in the early 1980s, leading to a 1989 Nobel award to Thomas Cech and Sidney Altman. In 1990, it was found in *Petunia* that introduced genes can silence similar genes of the plant's own, now known to be a result of RNA interference.

At about the same time, 22 nt long RNAs, now called microRNAs, were found to have a role in the development of *C. elegans*. Studies on RNA interference gleaned a Nobel Prize for Andrew Fire and Craig Mello in 2006, and another Nobel was awarded for studies on the transcription of RNA to Roger Kornberg in the same year. The discovery of gene regulatory RNAs has led to attempts to develop drugs made of RNA, such as siRNA, to silence genes. Adding to the Nobel prizes awarded for research on RNA in 2009 it was awarded for the elucidation of the atomic structure of the ribosome to Venki Ramakrishnan, Tom Steitz, and Ada Yonath.

Relevance for Prebiotic Chemistry and Abiogenesis

In 1967, Carl Woese hypothesized that RNA might be catalytic and suggested that the earliest forms of life (self-replicating molecules) could have relied on RNA both to carry genetic information and to catalyze biochemical reactions—an RNA world.

In March 2015, complex DNA and RNA nucleotides, including uracil, cytosine and thymine, were reportedly formed in the laboratory under outer space conditions, using starter chemicals, such as pyrimidine, an organic compound commonly found in meteorites. Pyrimidine, like polycyclic aromatic hydrocarbons (PAHs), is one of the most carbon-rich compounds found in the Universe and may have been formed in red giants or in interstellar dust and gas clouds.

References

- Klein, donald w.; lansing m.; harley, john (2006). Microbiology (6th ed.). New york: mcgraw-hill. Isbn 978-0-07-255678-0

- Gene, science: britannica.com, retrieved 22 May, 2019

- "allele noun - definition, pictures, pronunciation and usage notes". Oxford advanced learner›s dictionary. Retrieved 29 october 2017

- Cell-biology, science: britannica.com, retrieved 2 June, 2019

- "allele meaning in the cambridge english dictionary". Dictionary.cambridge.org. Retrieved 29 october 2017

- Gutteridge a, thornton jm (november 2005). "understanding nature›s catalytic toolkit". Trends in biochemical sciences. 30 (11): 622–29. Doi:10.1016/j.tibs.2005.09.006. Pmid 16214343

- Plasmids: microscopemaster.com, retrieved 23 February, 2019

- Hui, y. H. (2004). Handbook of vegetable preservation and processing. New york: m. Dekker. P. 180. Isbn 978-0-8247-4301-7. Oclc 52942889

- "human genome project: sequencing the human genome | learn science at scitable". Www.nature.com. Retrieved 2016-01-25

3
Fundamental Techniques in Biotechnology

Various techniques and methods are used in biotechnology. A few of them are cell culture, tissue culture, cloning, selective breeding, bioremediation and biosynthesis. The chapter closely examines these fundamental techniques in biotechnology to provide an extensive understanding of the subject.

Cell Culture

Cell culture refers to the removal of cells from an animal or plant and their subsequent growth in a favorable artificial environment. The cells may be removed from the tissue directly and disaggregated by enzymatic or mechanical means before cultivation, or they may be derived from a cell line or cell strain that has already been established.

Essentially, cell culture involves the distribution of cells in an artificial environment (in vitro) which is composed of the necessary nutrients, ideal temperature, gases, pH and humidity to allow the cells to grow and proliferate.

- In vivo - When the study involves living biological entities within the organism.

- In vitro - When the study is conducted using biological entities (cells, tissue etc) that has been isolated from their natural biological environment. E.g. tissue or cells isolated from the liver or kidney.

Whereas pieces of tissue can be put in the appropriate culture to produce cells that can then be used for culture (explant culture), cells from tissues (soft tissue) can be obtained through enzymatic reactions. Here, such enzymes as trypsin and proname are used to break down the tissue and release the desired cells.

When cells have been obtained directly from the organism/animal tissue (or even plant tissue) through enzymatic or mechanical techniques, such cells are referred to as primary cells. However, cells that continue to proliferate indefinitely (after the first subculture) under special conditions are referred to as cell lines.

These particular cells tend to have been passaged for a long period of time, which causes them to acquire homogenous (similar) genotypic and phenotypical traits.

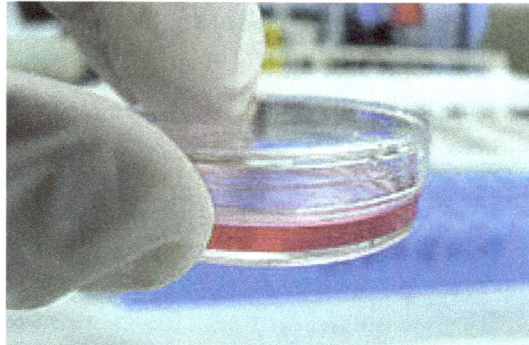

Cell Culture in Petri Dish.

Morphology

Based on their appearance, cells in culture can be categorized in to three main groups:

- Fibroblastic - This includes cells that tend to be bipolar/multipolar with elongated shapes. These cells are attached to the substrate as they grow.

- Epithelial - Epithelial like cells attain a polygonal shape with regular dimensions. Although they tend to grow in discrete patches, these cells also grow attached to the substrate.

- Lymphoblast - These cells are usually spherical in shape and do not attach to the surface of the substrate. As a result, they are grown in suspension.

For solid plates, solidifying agents (such as agar) are used where the liquid media is used with agar.

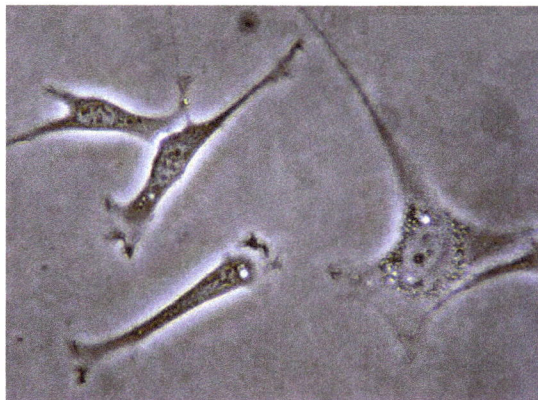

Significance of Cell Culture

Cell culture is an important technique in both cellular and molecular biology given that it provides the best platform for studying the normal physiology and biochemistry of cells. A cell is the basic structural, functional and biological unit of all living things.

In order to understand an organism or given tissues, it is important to understand how its cells work. Through cell culture, this becomes possible especially due to the fact the primary cells resemble the parental cells from the organism/tissue.

Whatever is learnt about the cells in vitro is representative of what is happening to the organism/ tissue. This makes cell culture significantly important for vaccine development, screening (drugs etc) and diagnosis of given diseases/conditions.

Given that different types of cells require different environments for proliferation, there are different types of media used for culture such as serum-free media and serum containing media among others.

Once the right requirements have been provided, the cells will increase in numbers and may form colonies, which can then be easily seen and identified. However, all this requires that the purpose of the procedure be understood.

Having a good understanding of what the procedure is meant to achieve, it becomes easier to prepare the culture with the right components. For instance, by understanding what the procedure is aimed for, the researcher will know whether to prepare a selective media (which allow for specific cells to grow) or differential media (allowing for different types of cells to grow).

Primary Cell Culture

Cell culture is a process where cells (animal or plant cells) are removed from the organism and introduced in to an artificial environment with favorable conditions for growth. This allows for researchers to study and learn more about the cells.

There are three major types of cell culture, which include:

- Primary cell culture,

- Secondary cell culture,

- Cell line.

Here, we shall focus on primary cell culture.

There are two types of primary cells:

- Adherent cells - Also referred to as anchorage dependent cells, these are the type of cells that require attachment for growth. Adherent cells are immobile, and obtained from such organs as kidney.

- Suspension cells - These are the type of cells that do not require attachment in order to grow. They are therefore also referred to as anchorage independent cells, and include such cells as lymphocytes found in the blood system.

In primary cell culture, cells obtained from such parental tissues (living tissues) as the liver and kidney, are introduced into suitable media for growth. Once the cells have been obtained, they can either be cultured as explants culture, suspension or monolayer.

In primary cell culture, the cells must have been obtained from the parental/living tissue. That is, they are not from another culture process.

Before the cells are cultured, they are first subjected to enzymatic treatment for dissociation. However, has to be for a minimal amount of time to avoid damaging or killing the cells. Once single cells are obtained, they are then appropriately cultured in media to allow them to grow (divide) are reach the desired numbers.

Initially, the culture tends to be heterogeneous in that it is composed of different types of cells obtained from the tissue. Although this can be maintained through the in vitro process (in a culture in a suitable media) this would only be for a limited period of time. However, through the transformation process, the primary cells may be used for a long period of time, changing the culture over time. These cells are refers to as continuous cell lines.

However, primary cells are typically preferred over continuous cell lines because of the fact that they are more similar (physiologically) to in vivo cells (cells from the living tissue). In addition, continuous cell lines may undergo certain changes (phenotypic and genotypic changes) which would result in discrepancies during analysis. As such, they cannot be used to determine what is happening to the in vivo cells. It is for this reason that primary cells are preferred.

Given that the primary cells significantly resemble the cells obtained from living tissue, they are important for research purposes in that they can be used to study their functions, metabolic regulations, cell physiology, development, defects and conditions affecting the tissue of interest.

In addition, they are used for such purposes as vaccine production, genetic engineering drug screening as well as toxicity testing and prenatal diagnosis among others.

Cell Culture Media

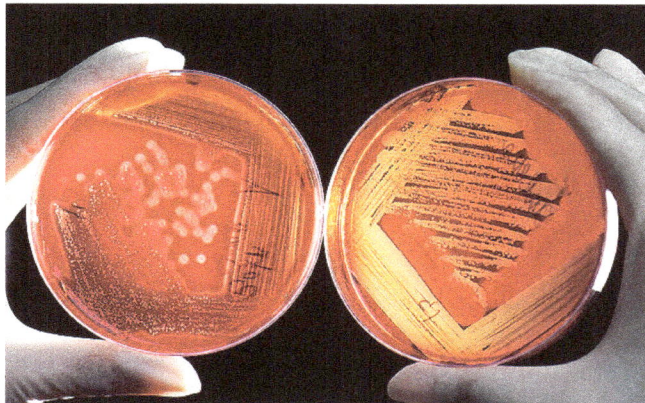

Blood agar plates are often used to diagnose infection. On the right is a positive Streptococcus culture; on the left is a positive Staphylococcus culture.

In cell culture techniques, cells (or tissues) are removed from a plant or an animal and introduced into a new, artificial environment that can support their proliferation (survival and growth).

Some of the requirements of such an environment for the proliferation of the cells include:

- A substrate (source of nutrition),

- Ideal temperature range (controlled),

- Growth medium,

- Ideal pH among others.

Here, we shall focus on the medium (cell culture media).

Although there are different types of culture media (for different types of cells) they are typically composed of:

- Glucose,

- Amino acids,

- Vitamins,

- Inorganic salts,

- Attachment factors etc.

There are two major types of culture media. These include:

- Natural media - Natural culture media is composed of biological fluids that are naturally occurring. Although this type of media can be used for a range of cells, its biggest disadvantage is that it may lack the exact components required by given cells, which can greatly affect reproducibility.

- Artificial media - Also referred to as synthetic media, artificial media refers to the type of media that is produced by adding such nutrients as vitamins, gases (oxygen and carbon dioxide) and protein among others. These organic and inorganic nutrients are added so as to meet the specific needs of given cells, and thus provide the ideal environment for their growth.

As such, they can be used for a number of purposes including:

- Providing immediate survival of the cells.

- Allowing for prolonged survival of the cells.

- Allowing for indefinite growth of the cells.

- Providing for specialized functions.

On the other hand, culture may be categorized as:

- Selective media - This is a special type of media that only allows for certain cells to grow. For instance, blood agar (used to isolate Streptococcus & Moraxella species) can be turned in to a selective media by adding antibiotics.

- Differential media - This type of media allows for different types of cells/microorganisms to grow depending on their metabolism.

Different types of synthetic media are prepared in a manner that will provide the ideal proliferation environment for given cells. For this reason, synthetic media can be divided in to four major categories.

These include:

- Serum containing media - In these types of media, serum (fetal bovine serum) is used as a carrier for nutrients and growth factors among others that tend to be water insoluble.

- Serum-free media - These types of media is typically produced for the purposes of supporting single cell type of culture. As such, it provides specified nutrients and other factors required by the cell type. In this media, serum is absent because it present some disadvantages and can result in misinterpretation of immunological results.

- Chemically defined media - Like the name suggests, this type of media is composed of contamination- free pure organic and inorganic ingredients. Constituents of this type of media are typically produced through genetic engineering in bacteria/yeast.

- Protein-free media - Protein- free media are typically lacking of any type of protein. It is largely used to promote superior growth of the cells as well as protein expression in addition to facilitating for the purification of any expressed product.

Some of the major components of cell culture media include:

- Nutrients - provided for by peptides and amino-acids, which are the building blocks of proteins.

- Carbohydrates for energy.

- Essential minerals such as calcium, magnesium, phosphates and iron among others buffering agents such s acetates to stabilize the culture media.

- Vitamins.

- PH change indicators such as phenol red.

Cell culture media are used for the proliferation of cells, which can then be identified and studied. As such, it can be used for various purposes including for education, diagnosis and treatment of a disease among others.

Cell Suspension

In culture methods, cell suspension refers to a type of culture where cells are suspended in a liquid medium.

To obtain single cells, a friable callus (small tissue that falls apart easily) is put in agitated liquid medium (agitation allows for gaseous exchange unlike solid medium), breaking it up. This allows for single cells to be released, which are then transferred to another fresh medium.

Cell suspension cultures have a big advantage over the stationary ones given that it allows for the cells to be uniformly bathed. Moreover, given that the medium tends to be agitated, it allows for aeration of the medium, providing gases to the cells. Given that the medium is a suspension, it also becomes easy to manipulate the contents of the culture.

Like any other culture, suspension cell culture has to be under controlled conditions, proving the cells with an ideal environment to proliferate. Once they reach about 80 percent confluence, it is time to subculture in order to ensure continued proper growth.

80 percent confluence refers to the state where 80 percent of the culture surface is covered with the growing cells.

In some cases, the cells in suspension may adhere on to the plastic surface of the culture flask or even form clumps. In such cases, a pipette can be used to pick these cells and expel them on to the surface of the flask and therefore away from the plastic surface. This helps obtain single cells given that they are no adhere on to the plastic surface.

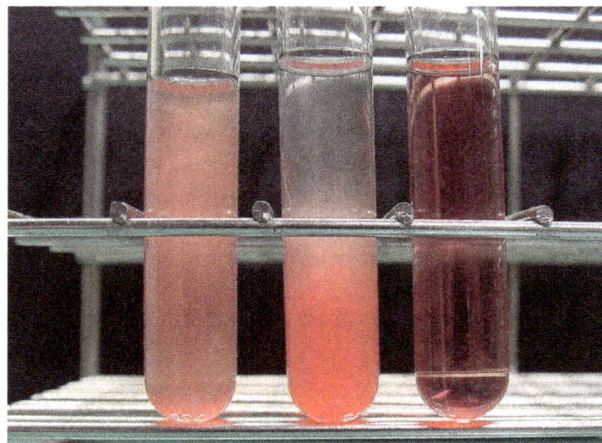

By Y tambe (Y tambe's file).

Red Blood Cell Suspension:

- (left: without hemolysis) Red blood cell suspension (0.5% sheep RBCs in saline), seems red and opaque.

- (middle: without hemolysis) RBCs sedimented spontaneously for 60 min. Note that the supernatant is not colored.

- (right: hemolysis) RBC suspension treated with the hemolysin of S. pyogenes at 37C for 30 min, become transparent by hemolysis.

Counting Cells

Counting the number of cells in a suspension is a process that involves the use of a stain. For instance, when trypan blue is used, it penetrates the cell membrane of the dead cells, but not the living cells.

The cells are then gently expelled into a haemocytometer (contains the counting chamber) under the cover slip and observed under a microscope. Cells are then counted within a given number of squares for calculations.

Significance

This method is largely preferred due to the fact that it allows for cells to be suspended in a solution rather than being held in a solid media. Here, therefore, it becomes easier to manipulate the contents thereby preventing them from forming clusters. With cell suspensions, it is also easier to observe single cells under the microscope.

In this case, it becomes possible to not only study the structure of the cells, but also get to observe how well they have differentiated; viewing dead and living cells under the microscope.

Cell Culture Protocol

Cell culture protocols are meant to ensure that culture procedures are carried out to the required standards. This is not only meant to prevent the contamination of the cells, but to also ensure that the researchers themselves are protected from any form of contamination.

Moreover, the nature of the work is expected to conform to the appropriate ethical guidelines. Therefore, before anything else, it is essential to ensure that the entire procedure conforms with both medical-ethical and animal- experiment guidelines. This is because going against such legislation and guidelines can result in heavy penalties and even shutting down of the laboratory.

Before any work starts, carry out the following procedure:

- Ensure that the working are is sanitized (using 70 percent ethanol).

- Always use a new pair of gloves. If a pair of gloves has to be used for another cell culture procedure, they should be sanitized using 70 percent ethanol and allowed to air dry.

- Any equipment that had been taken out of the cabinet should also be sanitized to prevent any contamination.

- Such equipment as pipette, glass jars and plastics to be used for the procedure should be autoclaved.

Although there are a wide range of culture media for cells, it is important to keep in mind that cell cultures, and particularly primary cell cultures are easily prone to contamination in addition to the risk of containing undetected viruses. For this reason, all material should be handled as potentially infectious in order to avoid any infections.

In addition, for safety purposes, work on cell culture should be carried out in the appropriate laminar flow hood, where air is directed away from the researcher.

Protocols for Cell Culture Preparation

Always check the information on the container to ensure that the medium is appropriate for the cell to be cultured.

Once prepared, the cell culture should be maintained under the recommended temperature range.

Monitor the culture every 30- 48 hours and check for confluency (when cells completely cover the surface of the culture) - However, this is largely dependent on the type of cells.

Once the procedure is completed and the cells have been analyzed, the culture should be appropriately discarded. Here, it is important to take a lot of caution given that by this time, cells have already proliferated and increased in numbers. Moreover, there are high chances that the specimen has been contaminated, which increase the risks of causing infections to the researcher if not handled appropriately.

Disposal

- For such items as scalpels, glass slides and cover slips among others, de- contamination may involve autoclaving at 121 degrees Celsius and 15 psi (pressure) for at least 30 minutes. However, if they are to be discarded, it is important that they are put into a plastic bag and into the appropriate container to be incinerated later.

- With liquid waste, chemical disinfection is one of the best methods through which the waste product can be deactivated. Such chemicals as bleach can be used for this purpose before pouring the liquid down the sink drain.

- For solid waste products, they are collected into a plastic bag for incineration. On the other hand, they can first be autoclaved before being incinerated.

Tissue Culture

Tissue culture refers to a method in which fragments of a tissue (plant or animal tissue) are introduced into a new, artificial environment, where they continue to function or grow. While fragments of a tissue are often used, it is important to note that entire organs are also used for tissue culture purposes. Here, such growth media as broth and agar are used to facilitate the process.

While the term tissue culture may be used for both plant and animal tissues, plant tissue culture is the more specific term used for the culture of plant tissues in tissue culture.

Types of Tissue Culture

Seed Culture

Seed culture is the type of tissue culture that is primarily used for plants such as orchids. For this method, explants (tissue from the plant) are obtained from an in-vitro derived plant and introduced in to an artificial environment, where they get to proliferate. In the event that a plant material is used directly for this process, then it has to be sterilized to prevent tissue damage and ensure optimum regeneration.

Embryo Culture

Embryo culture is the type of tissue culture that involves the isolation of an embryo from a given organism for in vitro growth.

Embryo culture may involve the use of a mature of immature embryo. Whereas mature embryos for culture are essentially obtained from ripe seeds, immature embryo (embryo rescue) involves the use of immature embryos from unripe/hybrid seeds that failed to germinate. In doing so, the embryo is ultimately able to produce a viable plant.

For embryo culture, the ovule, seed or fruit from which the embryo is to be obtained is sterilized, and therefore the embryo does not have to be sterilized again. Salt sucrose may be used to provide the embryo with nutrients. The culture is enriched with organic or inorganic compounds, inorganic salts as well as growth regulators.

Tissue Culture.

Callus Culture

Callus - This is the term used to refer to unspecialized, unorganized and a dividing mass of cells. A callus is produced when explants (cells) are cultured in an appropriate medium - A good example of this is the tumor tissue that grows out of the wounds of differentiated tissues/organs.

In practice, callus culture involves the growth of a callus (composed of differentiated and non- differentiated cells), which is the followed by a procedure that induces organ differentiation.

For this type of tissue culture, the culture is often sustained on a gel medium, which is composed of agar and a mixture of given macro and micronutrients depending on the type of cells. Different types of basal salt mixtures such as murashige and skoog medium are also used in addition to vitamins to enhance growth.

Callus Culture.

Organ Culture

Organ culture is a type of tissue culture that involves isolating an organ for in vitro growth. Here, any organ plant can be used as an explant for the culture process (Shoot, root, leaf, and flower).

With organ culture, or as is with their various tissue components, the method is used for preserve their structure or functions, which allows the organ to still resemble and retain the characteristics they would have in vivo. Here, new growth (differentiated structures) continues given that the organ retains its physiological features. As such, an organ helps provide information on patterns of growth, differentiation as well as development.

There are number of methods that can be used for organ culture. These include:

- Plasma clot method - Here, the method involves the use of a clot that is composed of plasma and chick embryo extract (or any other extract) in a watch glass. This method is particularly used for the purposes of studying morphogenesis in embryonic organ rudiments and more recently for studying the actions of various hormones, vitamins and carcinogens of adult mammalian tissues.

- Raft method - For this method, the explant is placed on a raft of lens paper/rayon acetate and floated on a serum in a watch glass.

- Agar gel method - The medium used for this method is composed of a salt solution, serum as well as the embryo extract or a mixture of various amino acids and vitamin with 1 percent agar. The explant has to be subcultured every 5 to 7 days. The method is largely used for the study of developmental aspects of normal organs and tumors.

- Grid method - Grid method, as the name suggests involves the use of perforated stainless steel sheet, on which the tissue of interest is placed before being placed in a culture chamber containing fluid medium.

Protoplast Culture

Protoplast - cells without cell walls. A protoplast is the term used to refer to cell (fungi, bacteria, plant cells etc) in which the cell wall has been removed, which is why they are also referred to as naked cells.

Protoplasts may be cultured in the following ways:

- Hanging-drop cultures,

- Micro culture chambers,

- Soft agars matrix.

Once a protoplast has regenerated a cell wall, then it goes through the process of cell division to form a callus, which may then be subcultured for continued growth.

Protoplast culture is an important method that provides numerous cells (single cells) that can be used for various studies. These include:

- Protoplast culture regenerated into a whole plant,

- Development of hybrids,

- Cell cloning,

- Genetic transformations,

- Membrane studies.

In protoplast culture, a number of phases can be observed. These include:

- Development of a cell wall,

- Cell division,

- Continuous growth or regeneration to a whole plant.

For plants, some of the special requirements include:

- Less amounts of iron and zinc and no ammonium,

- Higher concentration of calcium,

- High auxin/kinetic ratio for cell division and high kinetin/auxin ration for regeneration,

- Glucose and vitamins.

Some of the other types of tissue culture include:

- Single cell culture,

- Suspension culture,

- Anther culture,

- Pollen culture,

- Somatic Embryogenesis.

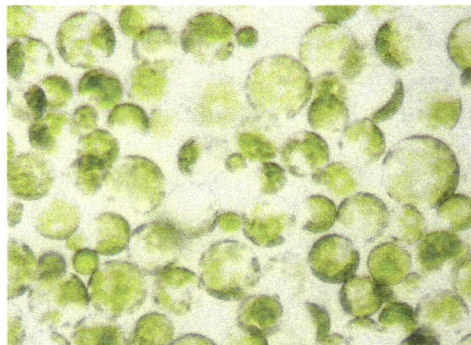

Protoplast Solution, Cells of Petunia leaves.

Major Steps of Tissue Culture (Plants)

Stage 1: Initiation Phase

The initiation phase is the first phase of tissue culture. Here, the tissue of interest is obtained and introduced and sterilized in order to prevent any microorganism from negatively affecting the process. It is during this stage that the tissue is initiated in to culture.

Stage 2: Multiplication Phase

The multiplication phase is the second step of tissue culture where the in vitro plant material is re- divided and then introduced in to the medium. Here, the medium is composed of appropriate components for growth including regulators and nutrients. These are responsible for the proliferation of the tissue and the production of multiple shoots.

This step is often repeated several times in order to obtain the desired number of plants.

Stage 3: Root Formation

It is at this phase that roots are formed. Here, hormones are required in order to induce rooting, and consequently complete plantlets.

Plant Tissue Culture

Tissue culture is applied in plant research for such purposes as the growing of new plants, which in some cases undergo genetic alterations. Here, the plant of interest is taken through the tissue culture process and grown in a controlled environment.

The Process of Plant Tissue Culture

This process involves the use of small pieces of a given plant tissue (plant of interest). Once the tissue is obtained, it is then cultured in the appropriate medium under sterile conditions so as to prevent various types of microorganisms from affecting the process.

The following is a general procedure for plant tissue culture:

Medium Preparation:

- The appropriate mixture (such as the MS mixture) is mixed with distilled water and stirred while adding the appropriate amount of sugar and sugar mixture. Here, sodium hydroxide or hydrochloric acid is used to adjust the pH - Contents used here will depend on the plant to be cultured and the number of tissues to be cultured.

- Agar is added to the mixture, heat and stirred to dissolve.

- After cooling, the warm medium is poured into polycarbonate tubes (to a depth of about 4 cm).

- With lids sitting on the tubes, the tubes are placed in a pressure cooker and sterilized for 20 minutes.

Plant preparation:

- Cut the plant part in to small pieces (e.g. cauliflower can be cut to florets of about 1cm across). On the other hand, such parts as the African violet leaves can be used as a whole.

- Using detergent and water, wash the plant part for about 20 minutes.

- Transfer the plant part in to sterilizing Clorox solution, shake for a minute and leave to sock for 20 minutes.

- Using a lid, gently discard the Clorox and retain the plant part in the container and then cap the container.

Transferring the plant material to a tissue culture medium 70 percent alcohol should be used for the sterilization of the equipment used and containers:

- Open the container and pour sterile water to cover half the container.

- Cover with a sterile lid again and shake the container for 2 to 3 minutes in order to wash the tissue and remove the bleach.

- Pour the water and repeat this three times.

- Using sterilized gloves, remove the plant part from the container and on to a sterile Petri dish.

- Using a sterile blade cut the plant material to smaller pieces of about 2 to 3 mm across avoiding the parts that have been damaged by bleach.

- Using sterile forceps, place a section of the plant in to the medium.

Cauliflower - partly submerged in medium with flower bud facing up.

Rose with shoots at level with medium surface.

African violet leaf laid directly in surface of medium depending on the plant used, it is important to check and find out how it should be placed in the medium.

Replace the lid/cap and close tightly:

This procedure will result in the development of a callus, which then produces shoots after a few weeks. Once the shoots develop, then the plant section may be placed in the right environment (well lit, warmth etc) for further growth.

Plant materials should be sterilized so as to remove any bacteria or spores that may be present.

For plants, the medium culture acts as a greenhouse that provides the explant with the idea environment for optimum growth. This includes being free of microorganisms, nutrients as well as the right balance of chemicals and hormones. Such media as BAP, TDZ are used while such hormones as IBA and IAA are used to induce growth. Some of the major reasons tissue culture is used for plants include:

- To produce large quantities of a given plant.

- To accelerate the production of new varieties of a plant.

- To maintain a virus free stock of the plant of interest.

Technique for Plant in Vitro Culture

- Micropropagation - This technique is used for the purposes of developing high- quality clonal plants (a clone is a group of identical cells). This has the potential to provide rapid and large scale propagation of new genotypes.

- Somatic cell genetics - Used for haploid production and somatic hybridization.

- Transgenic plants - Used for expression of mammalian genes or plant genes for various species it has proved beneficial for the engineering of species that are resistant against viruses and insects.

Biosynthesis

Biosynthesis is a multi-step, enzyme-catalyzed process where substrates are converted into more complex products in living organisms. In biosynthesis, simple compounds are modified, converted into other compounds, or joined together to form macromolecules. This process often consists of metabolic pathways. Some of these biosynthetic pathways are located within a single cellular organelle, while others involve enzymes that are located within multiple cellular organelles. Examples of these biosynthetic pathways include the production of lipid membrane components and nucleotides. Biosynthesis is usually synonymous with anabolism.

The prerequisite elements for biosynthesis include: precursor compounds, chemical energy (e.g. ATP), and catalytic enzymes which may require coenzymes (e.g.NADH, NADPH). These elements create monomers, the building blocks for macromolecules. Some important biological macromolecules include: proteins, which are composed of amino acid monomers joined via peptide bonds, and DNA molecules, which are composed of nucleotides joined via phosphodiester bonds.

Properties of Chemical Reactions

Biosynthesis occurs due to a series of chemical reactions. For these reactions to take place, the following elements are necessary:

- Precursor compounds: these compounds are the starting molecules or substrates in a reaction. These may also be viewed as the reactants in a given chemical process.

- Chemical energy: chemical energy can be found in the form of high energy molecules. These molecules are required for energetically unfavorable reactions. Furthermore, the hydrolysis of these compounds drives a reaction forward. High energy molecules, such as ATP, have three phosphates. Often, the terminal phosphate is split off during hydrolysis and transferred to another molecule.

- Catalytic enzymes: these molecules are special proteins that catalyze a reaction by increasing the rate of the reaction and lowering the activation energy.

- Coenzymes or cofactors: cofactors are molecules that assist in chemical reactions. These may be metal ions, vitamin derivatives such as NADH and acetyl CoA, or non-vitamin derivatives such as ATP. In the case of NADH, the molecule transfers a hydrogen, whereas acetyl CoA transfers an acetyl group, and ATP transfers a phosphate.

In the simplest sense, the reactions that occur in biosynthesis have the following format:

$$\text{Reactant} \xrightarrow{\text{enzyme}} \text{Product}$$

Some variations of this basic equation which will be discussed later in more detail are:

- Simple compounds which are converted into other compounds, usually as part of a multiple step reaction pathway. Two examples of this type of reaction occur during the formation of nucleic acids and the charging of tRNA prior to translation. For some of these steps, chemical energy is required:

$$\text{Precursor molecule} + \text{ATP} \rightleftharpoons \text{product AMP} + \text{PP}_i$$

- Simple compounds that are converted into other compounds with the assistance of cofactors. For example, the synthesis of phospholipids requires acetyl CoA, while the synthesis of another membrane component, sphingolipids, requires NADH and FADH for the formation the sphingosine backbone. The general equation for these examples is:

$$\text{Precursor molecule} + \text{Cofactor} \xrightarrow{\text{enzyme}} \text{Macromolecule}$$

- Simple compounds that join together to create a macromolecule. For example, fatty acids join together to form phospholipids. In turn, phospholipids and cholesterol interact noncovalently in order to form the lipid bilayer. This reaction may be depicted as follows:

$$\text{Molecule 1} + \text{Molecule 2} \rightarrow \text{macromolecule}$$

Lipid

Many intricate macromolecules are synthesized in a pattern of simple, repeated structures. For example, the simplest structures of lipids are fatty acids. Fatty acids are hydrocarbon derivatives; they contain a carboxyl group "head" and a hydrocarbon chain "tail". These fatty acids create larger components, which in turn incorporate noncovalent interactions to form the lipid bilayer. Fatty acid chains are found in two major components of membrane lipids: phospholipids and sphingolipids. A third major membrane component, cholesterol, does not contain these fatty acid units.

Phospholipids

The foundation of all biomembranes consists of a bilayer structure of phospholipids. The phospholipid molecule is amphipathic; it contains a hydrophilic polar head and a hydrophobic nonpolar tail. The phospholipid heads interact with each other and aqueous media, while the hydrocarbon tails orient themselves in the center, away from water. These latter interactions drive the bilayer structure that acts as a barrier for ions and molecules.

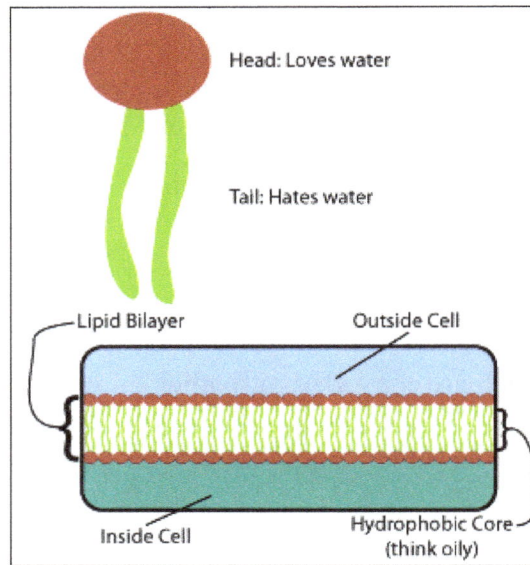

Lipid membrane bilayer.

There are various types of phospholipids; consequently, their synthesis pathways differ. However, the first step in phospholipid synthesis involves the formation of phosphatidate or diacylglycerol 3-phosphate at the endoplasmic reticulum and outer mitochondrial membrane. The synthesis pathway is found below:

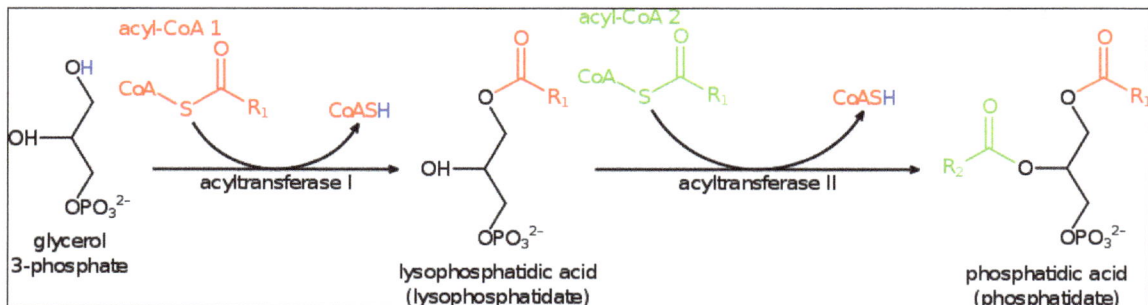

The pathway starts with glycerol 3-phosphate, which gets converted to lysophosphatidate via the addition of a fatty acid chain provided by acyl coenzyme A. Then, lysophosphatidate is converted to phosphatidate via the addition of another fatty acid chain contributed by a second acyl CoA; all of these steps are catalyzed by the glycerol phosphate acyltransferase enzyme. Phospholipid synthesis continues in the endoplasmic reticulum, and the biosynthesis pathway diverges depending on the components of the particular phospholipid.

Sphingolipids

Like phospholipids, these fatty acid derivatives have a polar head and nonpolar tails. Unlike phospholipids, sphingolipids have a sphingosine backbone. Sphingolipids exist in eukaryotic cells and are particularly abundant in the central nervous system. For example, sphingomyelin is part of the myelin sheath of nerve fibers.

Sphingolipids are formed from ceramides that consist of a fatty acid chain attached to the amino group of a sphingosine backbone. These ceramides are synthesized from the acylation of sphingosine. The biosynthetic pathway for sphingosine is found below:

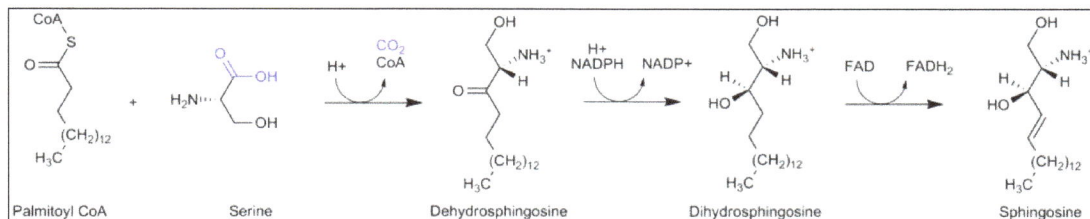

As the image denotes, during sphingosine synthesis, palmitoyl CoA and serine undergo a condensation reaction which results in the formation of dehydrosphingosine. This product is then reduced to form dihydrospingosine, which is converted to sphingosine via the oxidation reaction by FAD.

Cholesterol

This lipid belongs to a class of molecules called sterols. Sterols have four fused rings and a hydroxyl group. Cholesterol is a particularly important molecule. Not only does it serve as a component of lipid membranes, it is also a precursor to several steroid hormones, including cortisol, testosterone, and estrogen. Cholesterol is synthesized from acetyl CoA. The pathway is shown below:

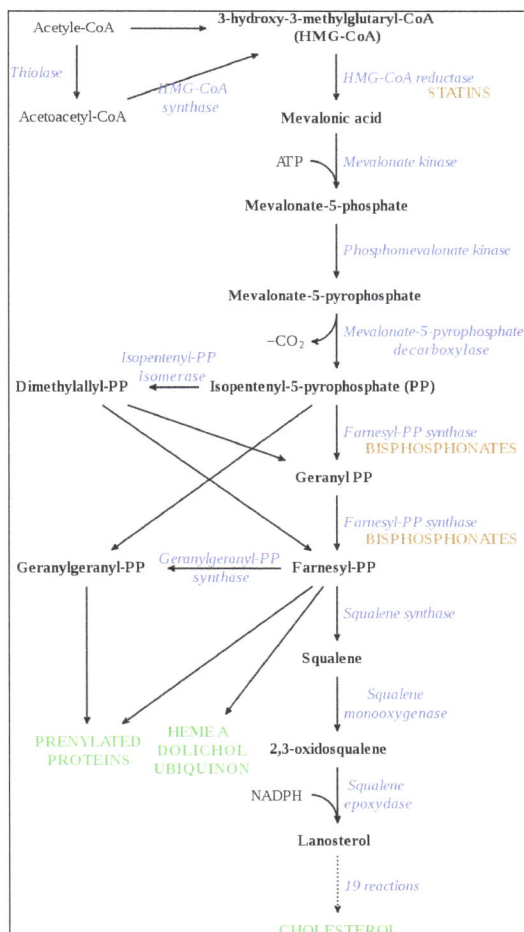

More generally, this synthesis occurs in three stages, with the first stage taking place in the cytoplasm and the second and third stages occurring in the endoplasmic reticulum. The stages are as follows:

- The synthesis of isopentenyl pyrophosphate, the "building block" of cholesterol.

- The formation of squalene via the condensation of six molecules of isopentenyl phosphate.

- The conversion of squalene into cholesterol via several enzymatic reactions.

Nucleotides

The biosynthesis of nucleotides involves enzyme-catalyzed reactions that convert substrates into more complex products. Nucleotides are the building blocks of DNA and RNA. Nucleotides are composed of a five-membered ring formed from ribose sugar in RNA, and deoxyribose sugar in DNA; these sugars are linked to a purine or pyrimidine base with a glycosidic bond and a phosphate group at the 5' location of the sugar.

Purine Nucleotides

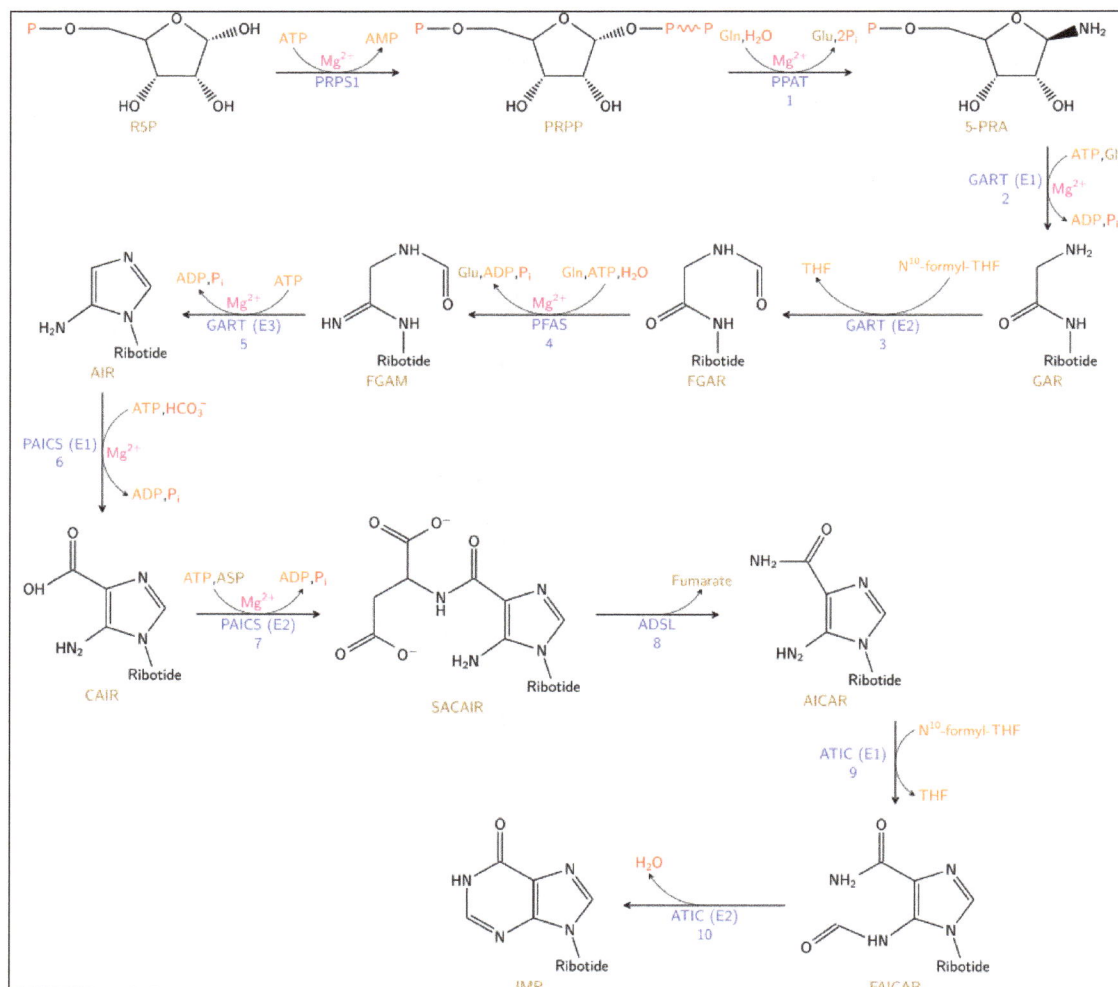

The synthesis of IMP.

The DNA nucleotides adenosine and guanosine consist of a purine base attached to a ribose sugar with a glycosidic bond. In the case of RNA nucleotides deoxyadenosine and deoxyguanosine, the purine bases are attached to a deoxyribose sugar with a glycosidic bond. The purine bases on DNA and RNA nucleotides are synthesized in a twelve-step reaction mechanism present in most single-celled organisms. Higher eukaryotes employ a similar reaction mechanism in ten reaction steps. Purine bases are synthesized by converting phosphoribosyl pyrophosphate (PRPP) to inosine monophosphate (IMP), which is the first key intermediate in purine base biosynthesis. Further enzymatic modification of IMP produces the adenosine and guanosine bases of nucleotides.

- The first step in purine biosynthesis is a condensation reaction, performed by glutamine-PRPP amidotransferase. This enzyme transfers the amino group from glutamine to PRPP, forming 5-phosphoribosylamine. The following step requires the activation of glycine by the addition of a phosphate group from ATP.

- GAR synthetase performs the condensation of activated glycine onto PRPP, forming glycineamide ribonucleotide (GAR).

- GAR transformylase adds a formyl group onto the amino group of GAR, forming formylglycinamide ribonucleotide (FGAR).

- FGAR amidotransferase catalyzes the addition of a nitrogen group to FGAR, forming formylglycinamidine ribonucleotide (FGAM).

- FGAM cyclase catalyzes ring closure, which involves removal of a water molecule, forming the 5-membered imidazole ring 5-aminoimidazole ribonucleotide (AIR).

- N5-CAIR synthetase transfers a carboxyl group, forming the intermediate N5-carboxyaminoimidazole ribonucleotide (N5-CAIR).

- N5-CAIR mutase rearranges the carboxyl functional group and transfers it onto the imidazole ring, forming carboxyamino- imidazole ribonucleotide (CAIR). The two step mechanism of CAIR formation from AIR is mostly found in single celled organisms. Higher eukaryotes contain the enzyme AIR carboxylase, which transfers a carboxyl group directly to AIR imidazole ring, forming CAIR.

- SAICAR synthetase forms a peptide bond between aspartate and the added carboxyl group of the imidazole ring, forming N-succinyl-5-aminoimidazole-4-carboxamide ribonucleotide (SAICAR).

- SAICAR lyase removes the carbon skeleton of the added aspartate, leaving the amino group and forming 5-aminoimidazole-4-carboxamide ribonucleotide (AICAR).

- AICAR transformylase transfers a carbonyl group to AICAR, forming N-formylaminoimidazole- 4-carboxamide ribonucleotide (FAICAR).

- The final step involves the enzyme IMP synthase, which performs the purine ring closure and forms the inosine monophosphate intermediate.

Pyrimidine Nucleotides

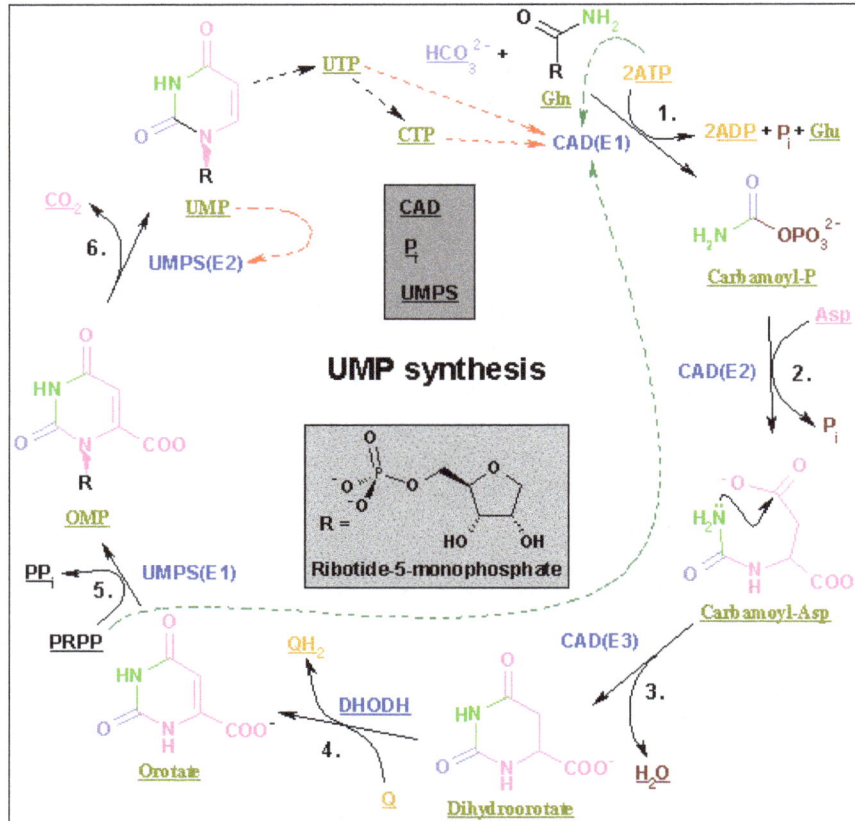

Uridine monophosphate (UMP) biosynthesis.

Other DNA and RNA nucleotide bases that are linked to the ribose sugar via a glycosidic bond are thymine, cytosine and uracil (which is only found in RNA). Uridine monophosphate biosynthesis involves an enzyme that is located in the mitochondrial inner membrane and multifunctional enzymes that are located in the cytosol.

1. The first step involves the enzyme carbamoyl phosphate synthase combining glutamine with CO_2 in an ATP dependent reaction to form carbamoyl phosphate.

2. Aspartate carbamoyltransferase condenses carbamoyl phosphate with aspartate to form uridosuccinate.

3. Dihydroorotase performs ring closure, a reaction that loses water, to form dihydroorotate.

4. Dihydroorotate dehydrogenase, located within the mitochondrial inner membrane, oxidizes dihydroorotate to orotate.

5. Orotate phosphoribosyl hydrolase (OMP pyrophosphorylase) condenses orotate with PRPP to form orotidine-5'-phosphate.

6. OMP decarboxylase catalyzes the conversion of orotidine-5'-phosphate to UMP.

After the uridine nucleotide base is synthesized, the other bases, cytosine and thymine are synthesized. Cytosine biosynthesis is a two-step reaction which involves the conversion of UMP to UTP. Phosphate addition to UMP is catalyzed by a kinase enzyme. The enzyme CTP synthase catalyzes the next reaction step: the conversion of UTP to CTP by transferring an amino group from glutamine to uridine; this forms the cytosine base of CTP. The mechanism, which depicts the reaction UTP + ATP + glutamine $\Leftrightarrow$ CTP + ADP + glutamate, is below:

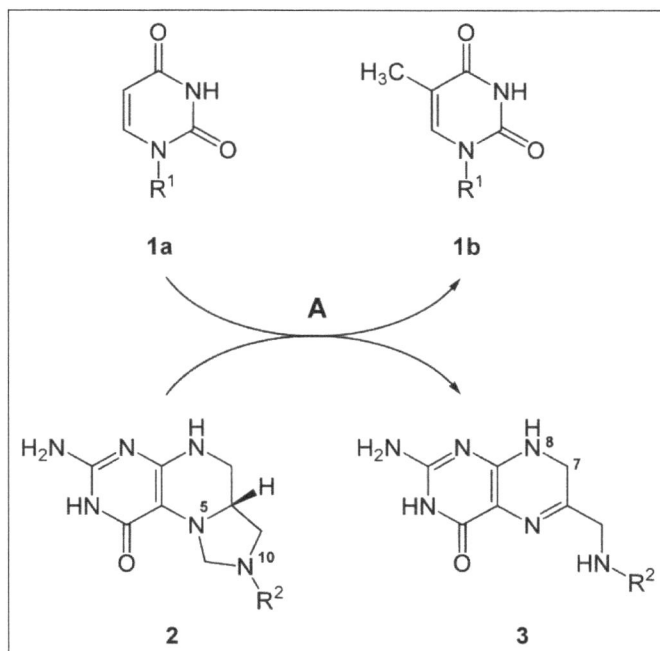

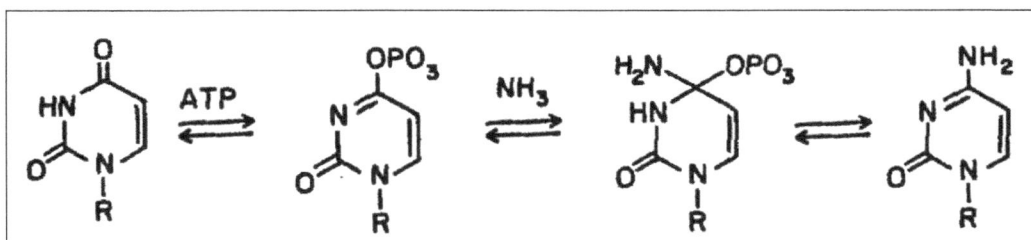

Cytosine is a nucleotide that is present in both DNA and RNA. However, uracil is only found in RNA. Therefore, after UTP is synthesized, it is must be converted into a deoxy form to be incorporated into DNA. This conversion involves the enzyme ribonucleoside triphosphate reductase. This reaction that removes the 2'-OH of the ribose sugar to generate deoxyribose is not affected by the bases attached to the sugar. This non-specificity allows ribonucleoside triphosphate reductase to convert all nucleotide triphosphates to deoxyribonucleotide by a similar mechanism.

In contrast to uracil, thymine bases are found mostly in DNA, not RNA. Cells do not normally contain thymine bases that are linked to ribose sugars in RNA, thus indicating that cells only synthesize deoxyribose-linked thymine. The enzyme thymidylate synthetase is responsible for synthesizing thymine residues from dUMP to dTMP. This reaction transfers a methyl group onto the uracil base of dUMP to generate dTMP. The thymidylate synthase reaction, dUMP + 5,10-methylenetetrahydrofolate $\Leftrightarrow$ dTMP + dihydrofolate, is shown to the right.

DNA

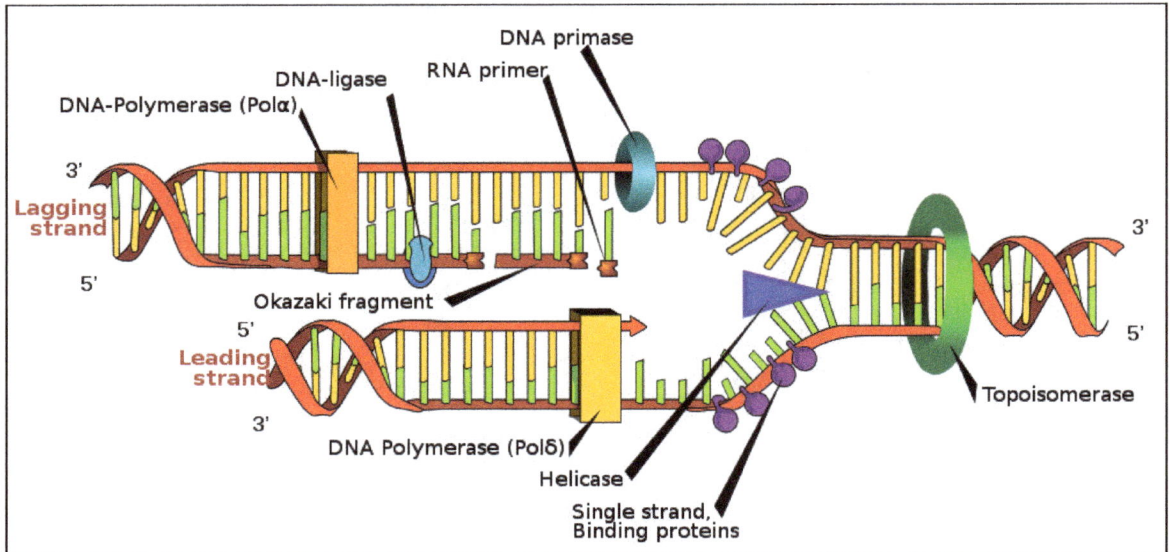

As DNA polymerase moves in a 3' to 5' direction along the template strand, it synthesizes
a new strand in the 5' to 3' direction.

DNA is composed of nucleotides that are joined by phosphodiester bonds. DNA synthesis, which takes place in the nucleus, is a semiconservative process, which means that the resulting DNA molecule contains an original strand from the parent structure and a new strand. DNA synthesis is catalyzed by a family of DNA polymerases that require four deoxynucleoside triphosphates, a template strand, and a primer with a free 3'OH in which to incorporate nucleotides.

In order for DNA replication to occur, a replication fork is created by enzymes called helicases which unwind the DNA helix. Topoisomerases at the replication fork remove supercoils caused by DNA unwinding, and single-stranded DNA binding proteins maintain the two single-stranded DNA templates stabilized prior to replication.

DNA synthesis is initiated by the RNA polymerase primase, which makes an RNA primer with a free 3'OH. This primer is attached to the single-stranded DNA template, and DNA polymerase elongates the chain by incorporating nucleotides; DNA polymerase also proofreads the newly synthesized DNA strand.

During the polymerization reaction catalyzed by DNA polymerase, a nucleophilic attack occurs by the 3'OH of the growing chain on the innermost phosphorus atom of a deoxynucleoside triphosphate; this yields the formation of a phosphodiester bridge that attaches a new nucleotide and releases pyrophosphate.

Two types of strands are created simultaneously during replication: the leading strand, which is synthesized continuously and grows towards the replication fork, and the lagging strand, which is made discontinuously in Okazaki fragments and grows away from the replication fork. Okazaki fragments are covalently joined by DNA ligase to form a continuous strand. Then, to complete DNA replication, RNA primers are removed, and the resulting gaps are replaced with DNA and joined via DNA ligase.

Amino Acids

A protein is a polymer that is composed from amino acids that are linked by peptide bonds. There are more than 300 amino acids found in nature of which only twenty, known as the standard amino acids, are the building blocks for protein. Only green plants and most microbes are able to synthesize all of the 20 standard amino acids that are needed by all living species. Mammals can only synthesize ten of the twenty standard amino acids. The other amino acids, valine, methionine, leucine, isoleucine, phenylalanine, lysine, threonine and tryptophan for adults and histidine, and arginine for babies are obtained through diet.

Amino Acid Basic Structure

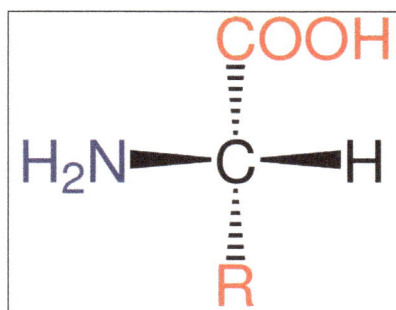

L-amino acid.

The general structure of the standard amino acids includes a primary amino group, a carboxyl group and the functional group attached to the α-carbon. The different amino acids are identified by the functional group. As a result of the three different groups attached to the α-carbon, amino acids are asymmetrical molecules. For all standard amino acids, except glycine, the α-carbon is a chiral center. In the case of glycine, the α-carbon has two hydrogen atoms, thus adding symmetry to this molecule. With the exception of proline, all of the amino acids found in life have the L-isoform conformation. Proline has a functional group on the α-carbon that forms a ring with the amino group.

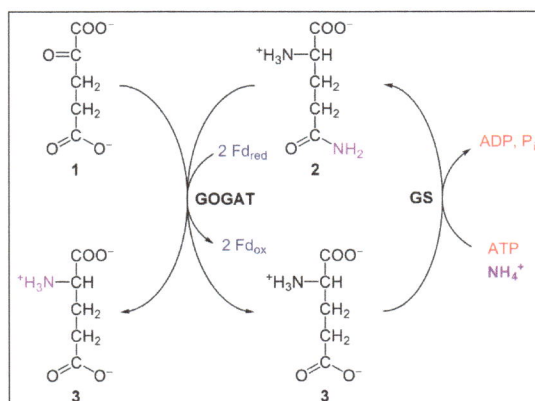

Nitrogen Source

One major step in amino acid biosynthesis involves incorporating a nitrogen group onto the α-carbon. In cells, there are two major pathways of incorporating nitrogen groups. One pathway involves the enzyme glutamine oxoglutarate aminotransferase (GOGAT) which removes the amide amino group of glutamine and transfers it onto 2-oxoglutarate, producing two glutamate molecules. In this catalysis reaction, glutamine serves as the nitrogen source. An image illustrating this reaction is found to the right.

The other pathway for incorporating nitrogen onto the α-carbon of amino acids involves the enzyme glutamate dehydrogenase (GDH). GDH is able to transfer ammonia onto 2-oxoglutarate and form glutamate. Furthermore, the enzyme glutamine synthetase (GS) is able to transfer ammonia onto glutamate and synthesize glutamine, replenishing glutamine.

The Glutamate Family of Amino Acids

The glutamate family of amino acids includes the amino acids that derive from the amino acid glutamate. This family includes: glutamate, glutamine, proline, and arginine. This family also includes the amino acid lysine, which is derived from α-ketoglutarate.

The biosynthesis of glutamate and glutamine is a key step in the nitrogen assimilation. The enzymes GOGAT and GDH catalyze the nitrogen assimilation reactions.

In bacteria, the enzyme glutamate 5-kinase initiates the biosynthesis of proline by transferring a phosphate group from ATP onto glutamate. The next reaction is catalyzed by the enzyme pyrroline-5-carboxylate synthase (P5CS), which catalyzes the reduction of the Y-carboxyl group of L-glutamate 5-phosphate. This results in the formation of glutamate semialdehyde, which spontaneously cyclizes to pyrroline-5-carboxylate. Pyrroline-5-carboxylate is further reduced by the enzyme pyrroline-5-carboxylate reductase (P5CR) to yield a proline amino acid.

In the first step of arginine biosynthesis in bacteria, glutamate is acetylated by transferring the acetyl group from acetyl-CoA at the N-α position; this prevents spontaneous cyclization. The enzyme N-acetylglutamate synthase (glutamate N-acetyltransferase) is responsible for catalyzing the acetylation step. Subsequent steps are catalyzed by the enzymes N-acetylglutamate kinase, N-acetyl-gamma-glutamyl-phosphate reductase, and acetylornithine/succinyldiamino pimelate aminotransferase and yield the N-acetyl-L-ornithine. The acetyl group of acetylornithine is removed by the enzyme acetylornithinase (AO) or ornithine acetyltransferase (OAT), and this yields ornithine. Then, the enzymes citrulline and argininosuccinate convert ornithine to arginine.

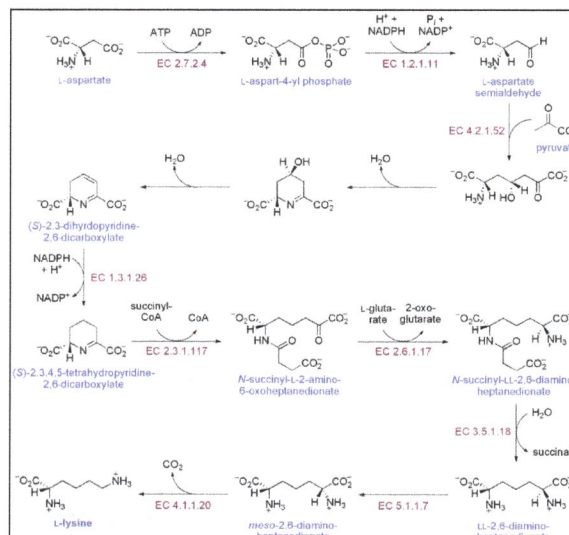

The diaminopimelic acid pathway.

There are two distinct lysine biosynthetic pathways: the diaminopimelic acid pathway and the α-aminoadipate pathway. The most common of the two synthetic pathways is the diaminopimelic acid pathway; it consists of several enzymatic reactions that add carbon groups to aspartate to yield lysine:

- Aspartate kinase initiates the diaminopimelic acid pathway by phosphorylating aspartate and producing aspartyl phosphate.

- Aspartate semialdehyde dehydrogenase catalyzes the NADPH-dependent reduction of aspartyl phosphate to yield aspartate semialdehyde.

- 4-hydroxy-tetrahydrodipicolinate synthase adds a pyruvate group to the β-aspartyl-4-semialdehyde, and a water molecule is removed. This causes cyclization and gives rise to (2S,4S)-4-hydroxy-2,3,4,5-tetrahydrodipicolinate.

- 4-hydroxy-tetrahydrodipicolinate reductase catalyzes the reduction of (2S,4S)-4-hydroxy-2,3,4,5-tetrahydrodipicolinate by NADPH to yield Δ'-piperideine-2,6-dicarboxylate (2,3,4,5-tetrahydrodipicolinate) and H_2O.

- Tetrahydrodipicolinate acyltransferase catalyzes the acetylation reaction that results in ring opening and yields N-acetyl α-amino-ε-ketopimelate.

- N-succinyl-α-amino-ε-ketopimelate-glutamate aminotransaminase catalyzes the transamination reaction that removes the keto group of N-acetyl α-amino-ε-ketopimelate and replaces it with an amino group to yield N-succinyl-L-diaminopimelate.

- N-acyldiaminopimelate deacylase catalyzes the deacylation of N-succinyl-L-diaminopimelate to yield L,L-diaminopimelate.

- DAP epimerase catalyzes the conversion of L,L-diaminopimelate to the meso form of L,L-diaminopimelate.

- DAP decarboxylase catalyzes the removal of the carboxyl group, yielding L-lysine.

The Serine Family of Amino Acids

The serine family of amino acid includes: serine, cysteine, and glycine. Most microorganisms and plants obtain the sulfur for synthesizing methionine from the amino acid cysteine. Furthermore, the conversion of serine to glycine provides the carbons needed for the biosynthesis of the methionine and histidine.

During serine biosynthesis, the enzyme phosphoglycerate dehydrogenase catalyzes the initial reaction that oxidizes 3-phospho-D-glycerate to yield 3-phosphonooxypyruvate. The following reaction is catalyzed by the enzyme phosphoserine aminotransferase, which transfers an amino group from glutamate onto 3-phosphonooxypyruvate to yield L-phosphoserine. The final step is catalyzed by the enzyme phosphoserine phosphatase, which dephosphorylates L-phosphoserine to yield L-serine.

There are two known pathways for the biosynthesis of glycine. Organisms that use ethanol and acetate as the major carbon source utilize the glyconeogenic pathway to synthesize glycine. The other pathway of glycine biosynthesis is known as the glycolytic pathway. This pathway converts serine synthesized from the intermediates of glycolysis to glycine. In the glycolytic pathway, the enzyme serine hydroxymethyltransferase catalyzes the cleavage of serine to yield glycine and transfers the cleaved carbon group of serine onto tetrahydrofolate, forming 5,10-methylene-tetrahydrofolate.

Cysteine biosynthesis is a two-step reaction that involves the incorporation of inorganic sulfur. In microorganisms and plants, the enzyme serine acetyltransferase catalyzes the transfer of acetyl group from acetyl-CoA onto L-serine to yield O-acetyl-L-serine. The following reaction step, catalyzed by the enzyme O-acetyl serine (thiol) lyase, replaces the acetyl group of O-acetyl-L-serine with sulfide to yield cysteine.

The Aspartate Family of Amino Acids

The aspartate family of amino acids includes: threonine, lysine, methionine, isoleucine, and aspartate. Lysine and isoleucine are considered part of the aspartate family even though part of their carbon skeleton is derived from pyruvate. In the case of methionine, the methyl carbon is derived from serine and the sulfur group, but in most organisms, it is derived from cysteine.

The biosynthesis of aspartate is a one step reaction that is catalyzed by a single enzyme. The enzyme aspartate aminotransferase catalyzes the transfer of an amino group from aspartate onto α-ketoglutarate to yield glutamate and oxaloacetate. Asparagine is synthesized by an ATP-dependent addition of an amino group onto aspartate; asparagine synthetase catalyzes the addition of nitrogen from glutamine or soluble ammonia to aspartate to yield asparagine.

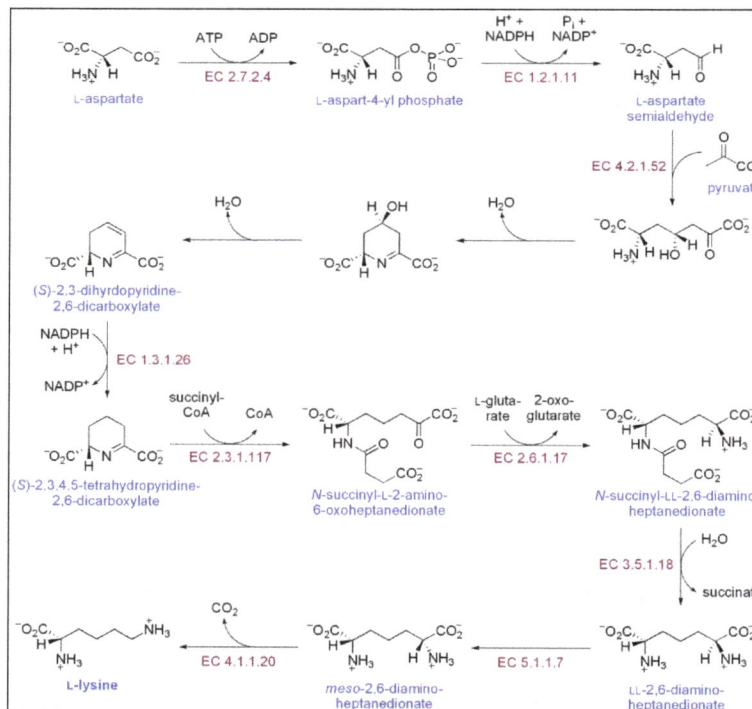

The diaminopimelic acid lysine biosynthetic pathway.

The diaminopimelic acid biosynthetic pathway of lysine belongs to the aspartate family of amino acids. This pathway involves nine enzyme-catalyzed reactions that convert aspartate to lysine.

- Aspartate kinase catalyzes the initial step in the diaminopimelic acid pathway by transferring a phosphoryl from ATP onto the carboxylate group of aspartate, which yields aspartyl-β-phosphate.

- Aspartate-semialdehyde dehydrogenase catalyzes the reduction reaction by dephosphorylation of aspartyl-β-phosphate to yield aspartate-β-semialdehyde.

- Dihydrodipicolinate synthase catalyzes the condensation reaction of aspartate-β-semialdehyde with pyruvate to yield dihydrodipicolinic acid.

- 4-hydroxy-tetrahydrodipicolinate reductase catalyzes the reduction of dihydrodipicolinic acid to yield tetrahydrodipicolinic acid.

- Tetrahydrodipicolinate N-succinyltransferase catalyzes the transfer of a succinyl group from succinyl-CoA on to tetrahydrodipicolinic acid to yield N-succinyl-L-2,6-diaminoheptanedioate.

- N-succinyldiaminopimelate aminotransferase catalyzes the transfer of an amino group from glutamate onto N-succinyl-L-2,6-diaminoheptanedioate to yield N-succinyl-L,L-diaminopimelic acid.

- Succinyl-diaminopimelate desuccinylase catalyzes the removal of acyl group from N-succinyl-L,L-diaminopimelic acid to yield L,L-diaminopimelic acid.

- Diaminopimelate epimerase catalyzes the inversion of the α-carbon of L,L-diaminopimelic acid to yield meso-diaminopimelic acid.

- Siaminopimelate decarboxylase catalyzes the final step in lysine biosynthesis that removes the carbon dioxide group from meso-diaminopimelic acid to yield L-lysine.

Proteins

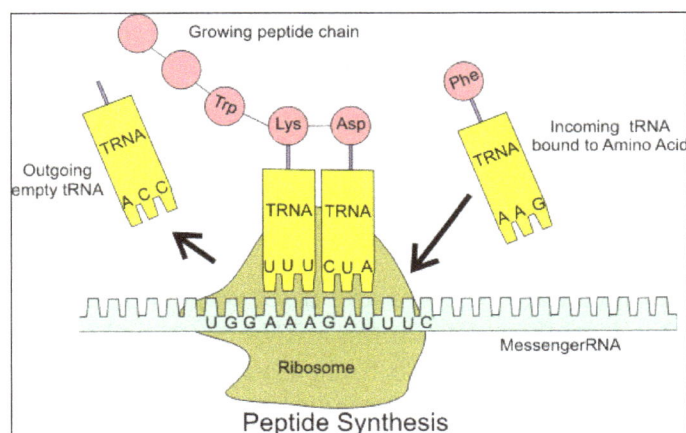

The tRNA anticodon interacts with the mRNA codon in order to bind an amino acid to growing polypeptide chain.

Protein synthesis occurs via a process called translation. During translation, genetic material called mRNA is read by ribosomes to generate a protein polypeptide chain. This process requires transfer RNA (tRNA) which serves as an adaptor by binding amino acids on one end and interacting with mRNA at the other end; the latter pairing between the tRNA and mRNA ensures that the correct amino acid is added to the chain. Protein synthesis occurs in three phases: initiation, elongation, and termination. Prokaryotic translation differs from eukaryotic translation; however, here we will mostly focus on the commonalities between the two organisms.

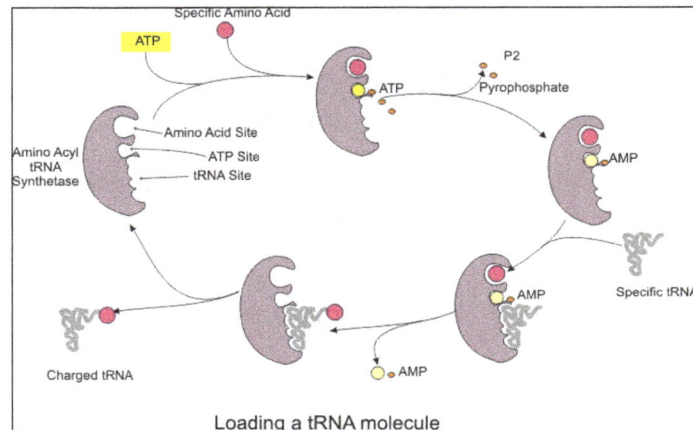

The process of tRNA charging.

Additional Background

Before translation can begin, the process of binding a specific amino acid to its corresponding tRNA must occur. This reaction, called tRNA charging, is catalyzed by aminoacyl tRNA synthetase. A specific tRNA synthetase is responsible for recognizing and charging a particular amino acid. Furthermore, this enzyme has special discriminator regions to ensure the correct binding between tRNA and its cognate amino acid. The first step for joining an amino acid to its corresponding tRNA is the formation of aminoacyl-AMP:

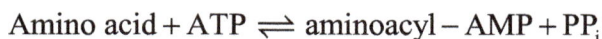

$$\text{Amino acid} + \text{ATP} \rightleftharpoons \text{aminoacyl} - \text{AMP} + \text{PP}_i$$

This is followed by the transfer of the aminoacyl group from aminoacyl-AMP to a tRNA molecule. The resulting molecule is aminoacyl-tRNA:

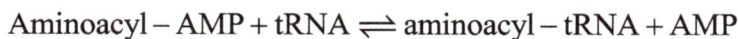

$$\text{Aminoacyl} - \text{AMP} + \text{tRNA} \rightleftharpoons \text{aminoacyl} - \text{tRNA} + \text{AMP}$$

The combination of these two steps, both of which are catalyzed by aminoacyl tRNA synthetase, produces a charged tRNA that is ready to add amino acids to the growing polypeptide chain.

In addition to binding an amino acid, tRNA has a three nucleotide unit called an anticodon that base pairs with specific nucleotide triplets on the mRNA called codons; codons encode a specific amino acid. This interaction is possible thanks to the ribosome, which serves as the site for protein synthesis. The ribosome possesses three tRNA binding sites: the aminoacyl site (A site), the peptidyl site (P site), and the exit site (E site).

There are numerous codons within an mRNA transcript, and it is very common for an amino acid to be specified by more than one codon; this phenomenon is called degeneracy. In all, there are 64 codons, 61 of each code for one of the 20 amino acids, while the remaining codons specify chain termination.

Translation in Steps

As previously mentioned, translation occurs in three phases: initiation, elongation, and termination.

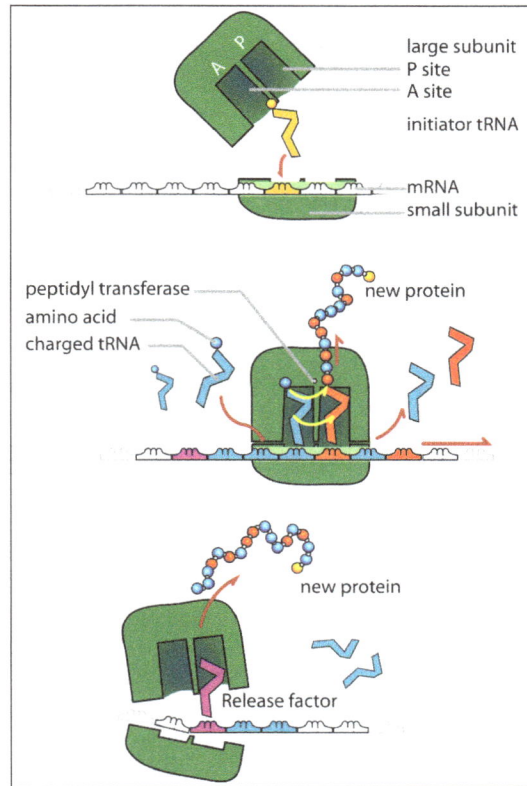

Translation.

Step 1: Initiation

The completion of the initiation phase is dependent on the following three events:

- The recruitment of the ribosome to mRNA.

- The binding of a charged initiator tRNA into the P site of the ribosome.

- The proper alignment of the ribosome with mRNA's start codon.

Step 2: Elongation

Following initiation, the polypeptide chain is extended via anticodon:codon interactions, with the ribosome adding amino acids to the polypeptide chain one at a time. The following steps must occur to ensure the correct addition of amino acids:

- The binding of the correct tRNA into the A site of the ribosome.

- The formation of a peptide bond between the tRNA in the A site and the polypeptide chain attached to the tRNA in the P site.

- Translocation or advancement of the tRNA-mRNA complex by three nucleotides.

Translocation "kicks off" the tRNA at the E site and shifts the tRNA from the A site into the P site, leaving the A site free for an incoming tRNA to add another amino acid.

Step 3: Termination

The last stage of translation occurs when a stop codon enters the A site. Then, the following steps occur:

- The recognition of codons by release factors, which causes the hydrolysis of the polypeptide chain from the tRNA located in the P site.

- The release of the polypeptide chain.

- The dissociation and "recycling" of the ribosome for future translation processes.

A summary table of the key players in translation is found below:

Key players in Translation	Translation Stage	Purpose
tRNA synthetase	before initiation	Responsible for tRNA charging
mRNA	initiation, elongation, termination	Template for protein synthesis; contains regions named codons which encode amino acids
tRNA	initiation, elongation, termination	Binds ribosomes sites A, P, E; anticodon base pairs with mRNA codon to ensure that the correct amino acid is incorporated into the growing polypeptide chain
ribosome	initiation, elongation, termination	Directs protein synthesis and catalyzes the formation of the peptide bond

Diseases Associated with Macromolecule Deficiency

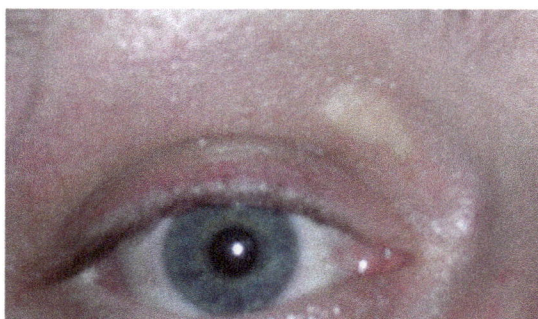

Familial hypercholesterolemia causes cholesterol deposits.

Errors in biosynthetic pathways can have deleterious consequences including the malformation of macromolecules or the underproduction of functional molecules. Below are examples that illustrate the disruptions that occur due to these inefficiencies.

- Familial hypercholesterolemia: this disorder is characterized by the absence of functional receptors for LDL. Deficiencies in the formation of LDL receptors may cause faulty receptors which disrupt the endocytic pathway, inhibiting the entry of LDL into the liver and other cells. This causes a buildup of LDL in the blood plasma, which results in atherosclerotic plaques that narrow arteries and increase the risk of heart attacks.

- Lesch–Nyhan syndrome: this genetic disease is characterized by self- mutilation, mental deficiency, and gout. It is caused by the absence of hypoxanthine-guanine phosphoribo-syltransferase, which is a necessary enzyme for purine nucleotide formation. The lack of enzyme reduces the level of necessary nucleotides and causes the accumulation of biosynthesis intermediates, which results in the aforementioned unusual behavior.

- Severe combined immunodeficiency (SCID): SCID is characterized by a loss of T cells. Shortage of these immune system components increases the susceptibility to infectious agents because the affected individuals cannot develop immunological memory. This immunological disorder results from a deficiency in adenosine deanimase activity, which causes a buildup of dATP. These dATP molecules then inhibit ribonucleotide reductase, which prevents of DNA synthesis.

- Huntington's disease: this neurological disease is caused from errors that occur during DNA synthesis. These errors or mutations lead to the expression of a mutant huntingtin protein, which contains repetitive glutamine residues that are encoded by expanding CAG trinucleotide repeats in the gene. Huntington's disease is characterized by neuronal loss and gliosis. Symptoms of the disease include: movement disorder, cognitive decline, and behavioral disorder.

Cloning

Cloning is the process of generating a genetically identical copy of a cell or an organism. Cloning happens all the time in nature—for example, when a cell replicates itself asexually without any genetic alteration or recombination. Prokaryotic organisms (organisms lacking a cell nucleus) such as bacteria create genetically identical duplicates of themselves using binary fission or budding. In eukaryotic organisms (organisms possessing a cell nucleus) such as humans, all the cells that undergo mitosis, such as skin cells and cells lining the gastrointestinal tract, are clones; the only exceptions are gametes (eggs and sperm), which undergo meiosis and genetic recombination.

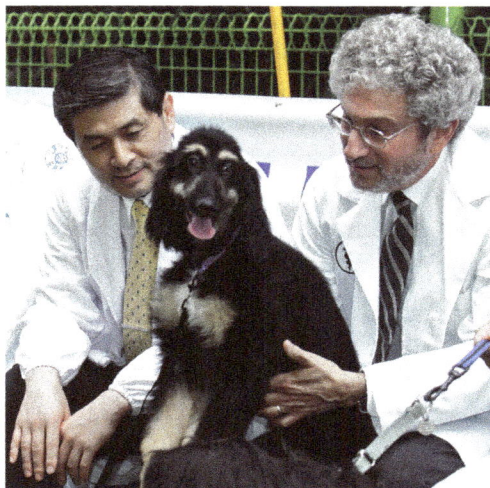

In biomedical research, cloning is broadly defined to mean the duplication of any kind of biological material for scientific study, such as a piece of DNA or an individual cell. For example, segments of DNA are replicated exponentially by a process known as polymerase chain reaction, or PCR, a technique that is used widely in basic biological research. The type of cloning that is the focus of much ethical controversy involves the generation of cloned embryos, particularly those of humans, which are genetically identical to the organisms from which they are derived, and the subsequent use of these embryos for research, therapeutic, or reproductive purposes.

Early Cloning Experiments

Reproductive cloning was originally carried out by artificial "twinning," or embryo splitting, which was first performed on a salamander embryo in the early 1900s by German embryologist Hans Spemann. Later, Spemann, who was awarded the Nobel Prize for Physiology or Medicine (1935) for his research on embryonic development, theorized about another cloning procedure known as nuclear transfer. This procedure was performed in 1952 by American scientists Robert W. Briggs and Thomas J. King, who used DNA from embryonic cells of the frog Rana pipiens to generate cloned tadpoles. In 1958 British biologist John Bertrand Gurdon successfully carried out nuclear transfer using DNA from adult intestinal cells of African clawed frogs (Xenopus laevis). Gurdon was awarded a share of the 2012 Nobel Prize in Physiology or Medicine for this breakthrough.

Advancements in the field of molecular biology led to the development of techniques that allowed scientists to manipulate cells and to detect chemical markers that signal changes within cells. With the advent of recombinant DNA technology in the 1970s, it became possible for scientists to create transgenic clones—clones with genomes containing pieces of DNA from other organisms. Beginning in the 1980s mammals such as sheep were cloned from early and partially differentiated embryonic cells. In 1996 British developmental biologist Ian Wilmut generated a cloned sheep, named Dolly, by means of nuclear transfer involving an enucleated embryo and a differentiated cell nucleus. This technique, which was later refined and became known as somatic cell nuclear transfer (SCNT), represented an extraordinary advance in the science of cloning, because it resulted in the creation of a genetically identical clone of an already grown sheep. It also indicated that it was possible for the DNA in differentiated somatic (body) cells to revert to an undifferentiated embryonic stage, thereby reestablishing pluripotency—the potential of an embryonic cell to grow into any one of the numerous different types of mature body cells that make up a complete organism. The realization that the DNA of somatic cells could be reprogrammed to a pluripotent state significantly impacted research into therapeutic cloning and the development of stem cell therapies.

Soon after the generation of Dolly, a number of other animals were cloned by SCNT, including pigs, goats, rats, mice, dogs, horses, and mules. Despite those successes, the birth of a viable SCNT primate clone would not come to fruition until 2018, and scientists used other cloning processes in the meantime. In 2001 a team of scientists cloned a rhesus monkey through a process called embryonic cell nuclear transfer, which is similar to SCNT except that it uses DNA from an undifferentiated embryo. In 2007 macaque monkey embryos were cloned by SCNT, but those clones lived only to the blastocyst stage of embryonic development. It was more than 10 years later, after improvements to SCNT had been made, that scientists announced the live birth of two clones of the crab-eating macaque (Macaca fascicularis), the first primate clones using the SCNT process. (SCNT has been carried out with very limited success in humans, in part because of problems with human egg cells resulting from the mother's age and environmental factors).

The first cloned cat, named CC (or Copy Cat), was born on December 22, 2001, to her surrogate mom, Allie (pictured).

Reproductive Cloning

Reproductive cloning involves the implantation of a cloned embryo into a real or an artificial uterus. The embryo develops into a fetus that is then carried to term. Reproductive cloning experiments were performed for more than 40 years through the process of embryo splitting, in which a single early-stage two-cell embryo is manually divided into two individual cells and then grows as two identical embryos. Reproductive cloning techniques underwent significant change in the 1990s, following the birth of Dolly, who was generated through the process of SCNT. This process entails the removal of the entire nucleus from a somatic (body) cell of an organism, followed by insertion of the nucleus into an egg cell that has had its own nucleus removed (enucleation). Once the somatic nucleus is inside the egg, the egg is stimulated with a mild electrical current and begins dividing. Thus, a cloned embryo, essentially an embryo of an identical twin of the original organism, is created. The SCNT process has undergone significant refinement since the 1990s, and procedures have been developed to prevent damage to eggs during nuclear extraction and somatic cell nuclear insertion. For example, the use of polarized light to visualize an egg cell's nucleus facilitates the extraction of the nucleus from the egg, resulting in a healthy, viable egg and thereby increasing the success rate of SCNT.

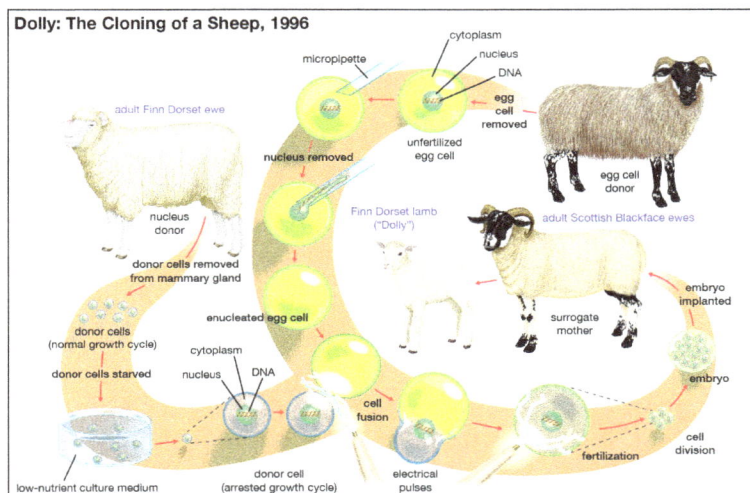

In figure, dolly the sheep was cloned using the process of somatic cell nuclear transfer (SCNT). While SCNT is used for cloning animals, it can also be used to generate embryonic stem cells. Prior to implantation of the fertilized egg into the uterus of the surrogate mother, the inner cell mass of the egg can be removed, and the cells can be grown in culture to form an embryonic stem cell line (generations of cells originating from the same group of parent cells).

Reproductive cloning using SCNT is considered very harmful since the fetuses of embryos cloned through SCNT rarely survive gestation and usually are born with birth defects. Wilmut's team of scientists needed 277 tries to create Dolly. Likewise, attempts to produce a macaque monkey clone in 2007 involved 100 cloned embryos, implanted into 50 female macaque monkeys, none of which gave rise to a viable pregnancy. In January 2008, scientists at Stemagen, a stem cell research and development company in California, announced that they had cloned five human embryos by means of SCNT and that the embryos had matured to the stage at which they could have been implanted in a womb. However, the scientists destroyed the embryos after five days, in the interest of performing molecular analyses on them.

Therapeutic Cloning

Therapeutic cloning is intended to use cloned embryos for the purpose of extracting stem cells from them, without ever implanting the embryos in a womb. Therapeutic cloning enables the cultivation of stem cells that are genetically identical to a patient. The stem cells could be stimulated to differentiate into any of the more than 200 cell types in the human body. The differentiated cells then could be transplanted into the patient to replace diseased or damaged cells without the risk of rejection by the immune system. These cells could be used to treat a variety of conditions, including Alzheimer disease, Parkinson disease, diabetes mellitus, stroke, and spinal cord injury. In addition, stem cells could be used for in vitro (laboratory) studies of normal and abnormal embryo development or for testing drugs to see if they are toxic or cause birth defects.

Although stem cells have been derived from the cloned embryos of animals such as mice, the generation of stem cells from cloned primate embryos has proved exceptionally difficult. For example, in 2007 stem cells successfully derived from cloned macaque embryos were able to differentiate into mature heart cells and brain neurons. However, the experiment started with 304 egg cells and resulted in the development of only two lines of stem cells, one of which had an abnormal Y chromosome. Likewise, the production of stem cells from human embryos has been fraught with the challenge of maintaining embryo viability. In 2001 scientists at Advanced Cell Technology, a research company in Massachusetts, successfully transferred DNA from human cumulus cells, which are cells that cling to and nourish human eggs, into eight enucleated eggs. Of these eight eggs, three developed into early-stage embryos (containing four to six cells); however, the embryos survived only long enough to divide once or twice. In 2004 South Korean researcher Hwang Woo Suk claimed to have cloned human embryos using SCNT and to have extracted stem cells from the embryos. However, this later proved to be a fraud; Hwang had fabricated evidence and had actually carried out the process of parthenogenesis, in which an unfertilized egg begins to divide with only half a genome. The following year a team of researchers from the University of Newcastle upon Tyne was able to grow a cloned human embryo to the 100-cell blastocyst stage using DNA from embryonic stem cells, though they did not generate a line of stem cells from the blastocyst. Scientists have since successfully derived embryonic stem cells from SCNT human embryos.

Progress in research on therapeutic cloning in humans has been slow relative to the advances made in reproductive cloning in animals. This is primarily because of the technical challenges and ethical controversy arising from the procuring of human eggs solely for research purposes. In addition, the development of induced pluripotent stem cells, which are derived from somatic cells that have been reprogrammed to an embryonic state through the introduction of specific genetic factors into the cell nuclei, has challenged the use of cloning methods and of human eggs.

Ethical Controversy

Human reproductive cloning remains universally condemned, primarily for the psychological, social, and physiological risks associated with cloning. A cloned embryo intended for implantation into a womb requires thorough molecular testing to fully determine whether an embryo is healthy and whether the cloning process is complete. In addition, as demonstrated by 100 failed attempts to generate a cloned macaque in 2007, a viable pregnancy is not guaranteed. Because the risks associated with reproductive cloning in humans introduce a very high likelihood of loss of life, the process is considered unethical. There are other philosophical issues that also have been raised concerning the nature of reproduction and human identity that reproductive cloning might violate. Concerns about eugenics, the once popular notion that the human species could be improved through the selection of individuals possessing desired traits, also have surfaced, since cloning could be used to breed "better" humans, thus violating principles of human dignity, freedom, and equality.

There also exists controversy over the ethics of therapeutic and research cloning. Some individuals and groups have an objection to therapeutic cloning, because it is considered the manufacture and destruction of a human life, even though that life has not developed past the embryonic stage. Those who are opposed to therapeutic cloning believe that the technique supports and encourages acceptance of the idea that human life can be created and expended for any purpose. However, those who support therapeutic cloning believe that there is a moral imperative to heal the sick and to seek greater scientific knowledge. Many of these supporters believe that therapeutic and research cloning should be not only allowed but also publicly funded, similar to other types of disease and therapeutics research. Most supporters also argue that the embryo demands special moral consideration, requiring regulation and oversight by funding agencies. In addition, it is important to many philosophers and policy makers that women and couples not be exploited for the purpose of obtaining their embryos or eggs.

There are laws and international conventions that attempt to uphold certain ethical principles and regulations concerning cloning. In 2005 the United Nations passed a nonbinding Declaration on Human Cloning that calls upon member states "to adopt all measures necessary to prohibit all forms of human cloning inasmuch as they are incompatible with human dignity and the protection of human life." This does provide leeway for member countries to pursue therapeutic cloning. The United Kingdom, through its Human Fertilisation and Embryology Authority, issues licenses for creating human embryonic stem cells through nuclear transfer. These licenses ensure that human embryos are cloned for legitimate therapeutic and research purposes aimed at obtaining scientific knowledge about disease and human development. The licenses require the destruction of embryos by the 14th day of development, since this is when embryos begin to develop the primitive streak, the first indicator of an organism's nervous system. The United States federal government has not passed any laws regarding human cloning due to disagreement within the legislative branch about whether to ban all cloning or to ban only reproductive cloning. The Dickey-Wicker amendment, attached to U.S. appropriations bills since 1995, has prevented the use of federal dollars to fund the harm or destruction of human embryos for research. It is presumed that nuclear transfer and any other form of cloning is subject to this restriction.

Selective Breeding

Selective breeding (also called artificial selection) is the process by which humans use animal breeding and plant breeding to selectively develop particular phenotypic traits (characteristics) by choosing which typically animal or plant males and females will sexually reproduce and have offspring together. Domesticated animals are known as breeds, normally bred by a professional breeder, while domesticated plants are known as varieties, cultigens, or cultivars. Two purebred animals of different breeds produce a crossbreed, and crossbred plants are called hybrids. Flowers, vegetables and fruit-trees may be bred by amateurs and commercial or non-commercial professionals: major crops are usually the provenance of the professionals.

In animal breeding, techniques such as inbreeding, linebreeding, and outcrossing are utilized. In plant breeding, similar methods are used. Charles Darwin discussed how selective breeding had been successful in producing change over time. He discussed selective breeding and domestication of such animals as pigeons, cats, cattle, and dogs. Darwin used artificial selection as a springboard to introduce and support the theory of natural selection.

The deliberate exploitation of selective breeding to produce desired results has become very common in agriculture and experimental biology.

Selective breeding can be unintentional, e.g., resulting from the process of human cultivation; and it may also produce unintended – desirable or undesirable – results. For example, in some grains, an increase in seed size may have resulted from certain ploughing practices rather than from the intentional selection of larger seeds. Most likely, there has been an interdependence between natural and artificial factors that have resulted in plant domestication.

A Belgian Blue cow. The defect in the breed's myostatin gene is maintained through linebreeding and is responsible for its accelerated lean muscle growth.

This Chihuahua mix and Great Dane shows the wide range of dog breed sizes created using selective breeding.

Animal Breeding

Three generations of "Westies" in a village.

Animals with homogeneous appearance, behavior, and other characteristics are known as particular breeds, and they are bred through culling animals with particular traits and selecting for further breeding those with other traits. Purebred animals have a single, recognizable breed, and purebreds with recorded lineage are called pedigreed. Crossbreeds are a mix of two purebreds, whereas mixed breeds are a mix of several breeds, often unknown. Animal breeding begins with breeding stock, a group of animals used for the purpose of planned breeding. When individuals are looking to breed animals, they look for certain valuable traits in purebred stock for a certain purpose, or may intend to use some type of crossbreeding to produce a new type of stock with different, and, it is presumed, superior abilities in a given area of endeavor. For example, to breed chickens, a breeder typically intends to receive eggs, meat, and new, young birds for further reproduction. Thus, the breeder has to study different breeds and types of chickens and analyze what can be expected from a certain set of characteristics before he or she starts breeding them. Therefore, when purchasing initial breeding stock, the breeder seeks a group of birds that will most closely fit the purpose intended.

Purebred breeding aims to establish and maintain stable traits, that animals will pass to the next generation. By "breeding the best to the best," employing a certain degree of inbreeding, considerable culling, and selection for "superior" qualities, one could develop a bloodline superior in certain respects to the original base stock. Such animals can be recorded with a breed registry, the organization that maintains pedigrees and/or stud books. However, single-trait breeding, breeding for only one trait over all others, can be problematic. In one case mentioned by animal behaviorist

Temple Grandin, roosters bred for fast growth or heavy muscles did not know how to perform typical rooster courtship dances, which alienated the roosters from hens and led the roosters to kill the hens after mating with them. A Soviet attempt to breed lab rats with higher intelligence led to cases of neurosis severe enough to make the animals incapable of any problem solving unless drugs like phenazepam were used.

The observable phenomenon of hybrid vigor stands in contrast to the notion of breed purity. However, on the other hand, indiscriminate breeding of crossbred or hybrid animals may also result in degradation of quality. Studies in evolutionary physiology, behavioral genetics, and other areas of organismal biology have also made use of deliberate selective breeding, though longer generation times and greater difficulty in breeding can make such projects challenging in vertebrates.

Plant Breeding

Researchers at the USDA have selectively bred carrots with a variety of colors.

Plant breeding has been used for thousands of years, and began with the domestication of wild plants into uniform and predictable agricultural cultigens. High-yielding varieties have been particularly important in agriculture.

Selective plant breeding is also used in research to produce transgenic animals that breed "true" (i.e., are homozygous) for artificially inserted or deleted genes.

Selective Breeding in Aquaculture

Selective breeding in aquaculture holds high potential for the genetic improvement of fish and shellfish. Unlike terrestrial livestock, the potential benefits of selective breeding in aquaculture were not realized until recently. This is because high mortality led to the selection of only a few broodstock, causing inbreeding depression, which then forced the use of wild broodstock. This was evident in selective breeding programs for growth rate, which resulted in slow growth and high mortality.

Control of the reproduction cycle was one of the main reasons as it is a requisite for selective breeding programs. Artificial reproduction was not achieved because of the difficulties in hatching or feeding some farmed species such as eel and yellowtail farming. A suspected reason associated with the late realisation of success in selective breeding programs in aquaculture was the education of the concerned people – researchers, advisory personnel and fish farmers. The education of fish biologists paid less attention to quantitative genetics and breeding plans.

Another was the failure of documentation of the genetic gains in successive generations. This in turn led to failure in quantifying economic benefits that successful selective breeding programs produce. Documentation of the genetic changes was considered important as they help in fine tuning further selection schemes.

Quality Traits in Aquaculture

Aquaculture species are reared for particular traits such as growth rate, survival rate, meat quality, resistance to diseases, age at sexual maturation, fecundity, shell traits like shell size, shell colour, etc.

- Growth rate – Growth rate is normally measured as either body weight or body length. This trait is of great economic importance for all aquaculture species as faster growth rate speeds up the turnover of production. Improved growth rates show that farmed animals utilize their feed more efficiently through a correlated response.

- Survival rate – Survival rate may take into account the degrees of resistance to diseases. This may also see the stress response as fish under stress are highly vulnerable to diseases. The stress fish experience could be of biological, chemical or environmental influence.

- Meat quality – The quality of fish is of great economic importance in the market. Fish quality usually takes into account size, meatiness, and percentage of fat, colour of flesh, taste, shape of the body, ideal oil and omega-3 content.

- Age at sexual maturation – The age of maturity in aquaculture species is another very important attribute for farmers as during early maturation the species divert all their energy to gonad production affecting growth and meat production and are more susceptible to health problems.

- Fecundity – As the fecundity in fish and shellfish is usually high it is not considered as a major trait for improvement. However, selective breeding practices may consider the size of the egg and correlate it with survival and early growth rate.

Finfish Response to Selection

Salmonids

Gjedrem showed that selection of Atlantic salmon (*Salmo salar*) led to an increase in body weight by 30% per generation. A comparative study on the performance of select Atlantic salmon with wild fish was conducted by AKVAFORSK Genetics Centre in Norway. The traits, for which the selection was done included growth rate, feed consumption, protein retention, energy retention, and feed conversion efficiency. Selected fish had a twice better growth rate, a 40% higher feed intake, and an increased protein and energy retention. This led to an overall 20% better Fed Conversion Efficiency as compared to the wild stock. Atlantic salmon have also been selected for resistance to bacterial and viral diseases. Selection was done to check resistance to Infectious

Pancreatic Necrosis Virus (IPNV). The results showed 66.6% mortality for low-resistant species whereas the high-resistant species showed 29.3% mortality compared to wild species.

Rainbow trout (*S. gairdneri*) was reported to show large improvements in growth rate after 7–10 generations of selection. Kincaid et al. showed that growth gains by 30% could be achieved by selectively breeding rainbow trout for three generations. A 7% increase in growth was recorded per generation for rainbow trout by Kause et al.

In Japan, high resistance to IPNV in rainbow trout has been achieved by selectively breeding the stock. Resistant strains were found to have an average mortality of 4.3% whereas 96.1% mortality was observed in a highly sensitive strain.

Coho salmon (*Oncorhynchus kisutch*) increase in weight was found to be more than 60% after four generations of selective breeding. In Chile, Neira et al. conducted experiments on early spawning dates in coho salmon. After selectively breeding the fish for four generations, spawning dates were 13–15 days earlier.

Cyprinids

Selective breeding programs for the Common carp (*Cyprinus carpio*) include improvement in growth, shape and resistance to disease. Experiments carried out in the USSR used crossings of broodstocks to increase genetic diversity and then selected the species for traits like growth rate, exterior traits and viability, and/or adaptation to environmental conditions like variations in temperature. Kirpichnikov *et al.* and Babouchkine selected carp for fast growth and tolerance to cold, the Ropsha carp. The results showed a 30–40% to 77.4% improvement of cold tolerance but did not provide any data for growth rate. An increase in growth rate was observed in the second generation in Vietnam. Moav and Wohlfarth showed positive results when selecting for slower growth for three generations compared to selecting for faster growth. Schaperclaus showed resistance to the dropsy disease wherein selected lines suffered low mortality (11.5%) compared to unselected (57%).

Channel Catfish

Growth was seen to increase by 12–20% in selectively bred *Iictalurus punctatus*. More recently, the response of the Channel Catfish to selection for improved growth rate was found to be approximately 80%, i.e., an average of 13% per generation.

Shellfish Response to Selection

Oysters

Selection for live weight of Pacific oysters showed improvements ranging from 0.4% to 25.6% compared to the wild stock. Sydney-rock oysters (*Saccostrea commercialis*) showed a 4% increase after one generation and a 15% increase after two generations. Chilean oysters (*Ostrea chilensis*), selected for improvement in live weight and shell length showed a 10–13% gain in one generation. Bonamia ostrea is a protistan parasite that causes catastrophic losses (nearly 98%) in European flat oyster *Ostrea edulis* L. This protistan parasite is endemic to three oyster-regions in Europe. Selective breeding programs show that *O. edulis* susceptibility to the infection differs across oyster strains in Europe. A study carried out by Culloty et al. showed that 'Rossmore' oysters in Cork harbour, Ireland had better resistance compared to other Irish strains. A selective breeding program at Cork harbour uses broodstock from 3– to 4-year-old survivors and is further controlled until a viable percentage reaches market size.

Over the years 'Rossmore' oysters have shown to develop lower prevalence of *B. ostreae* infection and percentage mortality. Ragone Calvo et al. selectively bred the eastern oyster, *Crassostrea virginica*, for resistance against co-occurring parasites *Haplosporidium nelson* (MSX) and *Perkinsus marinus* (Dermo). They achieved dual resistance to the disease in four generations of selective breeding. The oysters showed higher growth and survival rates and low susceptibility to the infections. At the end of the experiment, artificially selected *C. virginica* showed a 34–48% higher survival rate.

Penaeid Shrimps

Selection for growth in Penaeid shrimps yielded successful results. A selective breeding program for *Litopenaeus stylirostris* saw an 18% increase in growth after the fourth generation and 21% growth after the fifth generation. *Marsupenaeus japonicas* showed a 10.7% increase in growth after the first generation. Argue et al. conducted a selective breeding program on the Pacific White Shrimp, *Litopenaeus vannamei* at The Oceanic Institute, Waimanalo, USA from 1995 to 1998. They reported significant responses to selection compared to the unselected control shrimps. After one generation, a 21% increase was observed in growth and 18.4% increase in survival to TSV. The Taura Syndrome Virus (TSV) causes mortalities of 70% or more in shrimps. C.I. Oceanos S.A. in Colombia selected the survivors of the disease from infected ponds and used them as parents for the next generation. They achieved satisfying results in two or three generations wherein survival rates approached levels before the outbreak of the disease. The resulting heavy losses (up to 90%) caused by Infectious hypodermal and haematopoietic necrosis virus (IHHNV) caused a number of shrimp farming industries started to selectively breed shrimps resistant to this disease. Successful outcomes led to development of Super Shrimp, a selected line of *L. stylirostris* that is resistant to IHHNV infection. Tang et al. confirmed this by showing no mortalities in IHHNV- challenged Super Shrimp post larvae and juveniles.

Aquatic Species versus Terrestrial Livestock

Selective breeding programs for aquatic species provide better outcomes compared to terrestrial livestock. This higher response to selection of aquatic farmed species can be attributed to the following:

- High fecundity in both sexes fish and shellfish enabling higher selection intensity.

- Large phenotypic and genetic variation in the selected traits.

Selective breeding in aquaculture provide remarkable economic benefits to the industry, the primary one being that it reduces production costs due to faster turnover rates. This is because of faster growth rates, decreased maintenance rates, increased energy and protein retention, and better feed efficiency. Applying such genetic improvement program to aquaculture species will increase productivity to meet the increasing demands of growing populations.

Advantages and Disadvantages

Selective breeding is a direct way to determine if a specific trait can evolve in response to selection. A single-generation method of breeding is not as accurate or direct. The process is also more practical and easier to understand than sibling analysis. Selective breeding is better for traits such as physiology and behavior that are hard to measure because it requires fewer individuals to test than single-generation testing.

However, there are disadvantages to this process. Because a single experiment done in selective breeding cannot be used to assess an entire group of genetic variances, individual experiments must be done for every individual trait. Also, because of the necessity of selective breeding experiments to require maintaining the organisms tested in a lab or greenhouse, it is impractical to use this breeding method on many organisms. Controlled mating instances are difficult to carry out in this case and this is a necessary component of selective breeding.

Bioremediation

Bioremediation is a process used to treat contaminated media, including water, soil and subsurface material, by altering environmental conditions to stimulate growth of microorganisms and degrade the target pollutants. In many cases, bioremediation is less expensive and more sustainable than other remediation alternatives. Biological treatment is a similar approach used to treat wastes including wastewater, industrial waste and solid waste.

Most bioremediation processes involve oxidation-reduction reactions where either an electron acceptor (commonly oxygen) is added to stimulate oxidation of a reduced pollutant (e.g. hydrocarbons) or an electron donor (commonly an organic substrate) is added to reduce oxidized pollutants (nitrate, perchlorate, oxidized metals, chlorinated solvents, explosives and propellants). In both these approaches, additional nutrients, vitamins, minerals, and pH buffers may be added to optimize conditions for the microorganisms. In some cases, specialized microbial cultures are added (bioaugmentation) to further enhance biodegradation. Some examples of bioremediation related technologies are phytoremediation, mycoremediation, bioventing, bioleaching, landfarming, bioreactor, composting, bioaugmentation, rhizofiltration, and biostimulation.

Chemistry

Most bioremediation processes involve oxidation-reduction (Redox) reactions where a chemical species donates an electron (electron donor) to a different species that accepts the electron (electron acceptor). During this process, the electron donor is said to be oxidized while the electron acceptor is reduced. Common electron acceptors in bioremediation processes include oxygen, nitrate, manganese (III and IV), iron (III), sulfate, carbon dioxide and some pollutants (chlorinated solvents, explosives, oxidized metals, and radionuclides). Electron donors include sugars, fats, alcohols, natural organic material, fuel hydrocarbons and a variety of reduced organic pollutants. The redox potential for common biotransformation reactions is shown in the table.

Process	Reaction	Redox potential (E_h in mV)
aerobic	$O_2 + 4e^- + 4H^+ \rightarrow 2H_2O$	600 ~ 400
anaerobic		
denitrification	$2NO_3^- + 10e^- + 12H^+ \rightarrow N_2 + 6H_2O$	500 ~ 200
manganese IV reduction	$MnO_2 + 2e^- + 4H^+ \rightarrow Mn^{2+} + 2H_2O$	400 ~ 200
iron III reduction	$Fe(OH)_3 + e^- + 3H^+ \rightarrow Fe^{2+} + 3H_2O$	300 ~ 100
sulfate reduction	$SO_4^{2-} + 8e^- + 10 H^+ \rightarrow H_2S + 4H_2O$	0 ~ -150
fermentation	$2CH_2O \rightarrow CO_2 + CH_4$	-150 ~ -220

Aerobic

Aerobic bioremediation is the most common form of oxidative bioremediation process where oxygen is provided as the electron acceptor for oxidation of petroleum, polyaromatic hydrocarbons (PAHs), phenols, and other reduced pollutants. Oxygen is generally the preferred electron acceptor because of the higher energy yield and because oxygen is required for some enzyme systems to initiate the degradation process. Numerous laboratory and field studies have shown that microorganisms can degrade a wide variety of hydrocarbons, including components of gasoline, kerosene, diesel, and jet fuel. Under ideal conditions, the biodegradation rates of the low- to moderate-weight aliphatic, alicyclic, and aromatic compounds can be very high. As the molecular weight of the compound increases, so does the resistance to biodegradation.

Common approaches for providing oxygen above the water table include landfarming, composting and bioventing. During landfarming, contaminated soils, sediments, or sludges are incorporated into the soil surface and periodically turned over (tilled) using conventional agricultural equipment to aerate the mixture. Composting accelerates pollutant biodegradation by mixing the waste to be treated with a bulking agent, forming into piles, and periodically mixed to increase oxygen transfer. Bioventing is a process that increases the oxygen or air flow into the unsaturated zone of the soil which increases the rate of natural in situ degradation of the targeted hydrocarbon contaminant.

Approaches for oxygen addition below the water table include recirculating aerated water through the treatment zone, addition of pure oxygen or peroxides, and air sparging. Recirculation systems typically consist of a combination of injection wells or galleries and one or more recovery wells where the extracted groundwater is treated, oxygenated, amended with nutrients and reinjected. However, the amount of oxygen that can be provided by this method is limited by the low solubility of oxygen in water (8 to 10 mg/L for water in equilibrium with air at typical temperatures). Greater amounts of oxygen can be provided by contacting the water with pure oxygen or addition of hydrogen peroxide (H_2O_2) to the water. In some cases, slurries of solid calcium or magnesium peroxide are injected under pressure through soil borings. These solid peroxides react with water releasing H_2O_2 which then decomposes releasing oxygen. Air sparging involves the injection of air under pressure below the water table. The air injection pressure must be great enough to overcome the hydrostatic pressure of the water and resistance to air flow through the soil.

Anaerobic

Anaerobic bioremediation can be employed to treat a broad range of oxidized contaminants including chlorinated ethenes (PCE, TCE, DCE, VC), chlorinated ethanes (TCA, DCA), chloromethanes (CT, CF), chlorinated cyclic hydrocarbons, various energetics (e.g., perchlorate, RDX, TNT), and nitrate. This process involves the addition of an electron donor to: 1) deplete background electron acceptors including oxygen, nitrate, oxidized iron and manganese and sulfate; and 2) stimulate the biological and/or chemical reduction of the oxidized pollutants. Hexavalent chromium (Cr[VI]) and uranium (U[VI]) can be reduced to less mobile and/or less toxic forms (e.g., Cr[III], U[IV]). Similarly, reduction of sulfate to sulfide (sulfidogenesis) can be used to precipitate certain metals (e.g., zinc, cadmium). The choice of substrate and the method of injection depend on the contaminant type and distribution in the aquifer, hydrogeology, and remediation objectives. Substrate can be added using conventional well installations, by direct-push technology, or by excavation and backfill such as permeable reactive barriers (PRB) or biowalls. Slow-release products composed of edible oils or solid substrates tend to stay in place for an extended treatment period. Soluble substrates or soluble fermentation products of slow-release substrates can potentially migrate via advection and diffusion, providing broader but shorter-lived treatment zones. The added organic substrates are first fermented to hydrogen (H_2) and volatile fatty acids (VFAs). The VFAs, including acetate, lactate, propionate and butyrate, provide carbon and energy for bacterial metabolism.

Heavy Metals

Heavy metals including cadmium, chromium, lead and uranium are elements so they cannot be biodegraded. However, bioremediation processes can potentially be used to reduce the mobility of these material in the subsurface, reducing the potential for human and environmental exposure. The mobility of certain metals including chromium (Cr) and uranium (U) varies depending on the oxidation state of the material. Microorganisms can be used to reduce the toxicity and mobility of chromium by reducing hexavalent chromium, Cr(VI) to trivalent Cr (III). Uranium can be reduced from the more mobile U(VI) oxidation state to the less mobile U(IV) oxidation state. Microorganisms are used in this process because the reduction rate of these metals is often slow unless catalyzed by microbial interactions Research is also underway to develop methods to remove metals from water by enhancing the sorption of the metal to cell walls. This approach has been evaluated for treatment of cadmium, chromium, and lead.

Additives

In the event of biostimulation, adding nutrients that are limited to make the environment more suitable for bioremediation, nutrients such as nitrogen, phosphorus, oxygen, and carbon may be added to the system to improve effectiveness of the treatment.

Many biological processes are sensitive to pH and function most efficiently in near neutral conditions. Low pH can interfere with pH homeostasis or increase the solubility of toxic metals. Microorganisms can expend cellular energy to maintain homeostasis or cytoplasmic conditions may change in response to external changes in pH. Some anaerobes have adapted to low pH conditions through alterations in carbon and electron flow, cellular morphology, membrane structure, and protein synthesis.

Limitations

Bioremediation can be used to completely mineralize organic pollutants, to partially transform the pollutants, or alter their mobility. Heavy metals and radionuclides are elements that cannot be bio-degraded, but can be bio-transformed to less mobile forms. In some cases, microbes do not fully mineralize the pollutant, potentially producing a more toxic compound. For example, under anaerobic conditions, the reductive dehalogenation of TCE may produce dichloroethylene (DCE) and vinyl chloride (VC), which are suspected or known carcinogens. However, the microorganism *Dehalococcoides* can further reduce DCE and VC to the non-toxic product ethene. Additional research is required to develop methods to ensure that the products from biodegradation are less persistent and less toxic than the original contaminant. Thus, the metabolic and chemical pathways of the microorganisms of interest must be known. In addition, knowing these pathways will help develop new technologies that can deal with sites that have uneven distributions of a mixture of contaminants.

Also, for biodegradation to occur, there must be a microbial population with the metabolic capacity to degrade the pollutant, an environment with the right growing conditions for the microbes, and the right amount of nutrients and contaminants. The biological processes used by these microbes are highly specific, therefore, many environmental factors must be taken into account and regulated as well. Thus, bioremediation processes must be specifically made in accordance to the conditions at the contaminated site. Also, because many factors are interdependent, small-scale tests are usually performed before carrying out the procedure at the contaminated site. However, it can be difficult to extrapolate the results from the small-scale test studies into big field operations. In many cases, bioremediation takes more time than other alternatives such as land filling and incineration.

Genetic Engineering

The use of genetic engineering to create organisms specifically designed for bioremediation is under preliminary research. Two category of genes can be inserted in the organism: degradative genes which encode proteins required for the degradation of pollutants, and reporter genes that are able to monitor pollution levels. Numerous members of *Pseudomonas* have also been modified with the lux gene, but for the detection of the polyaromatic hydrocarbon naphthalene. A field test for the release of the modified organism has been successful on a moderately large scale.

There are concerns surrounding release and containment of genetically modified organisms into the environment due to the potential of horizontal gene transfer. Genetically modified organisms are classified and controlled under the Toxic Substances Control Act of 1976 under United States Environmental Protection Agency. Measures have been created to address these concerns. Organisms can be modified such that they can only survive and grow under specific sets of environmental conditions. In addition, the tracking of modified organisms can be made easier with the insertion of bioluminescence genes for visual identification.

Fermentation

Fermentation is a metabolic process that produces chemical changes in organic substrates through the action of enzymes. In biochemistry, it is narrowly defined as the extraction of energy from carbohydrates in the absence of oxygen. In the context of food production, it may more broadly refer to any process in which the activity of microorganisms brings about a desirable change to a foodstuff or beverage. The science of fermentation is known as zymology.

In microorganisms, fermentation is the primary means of producing ATP by the degradation of organic nutrients anaerobically. Humans have used fermentation to produce foodstuffs and beverages since the Neolithic age. For example, fermentation is used for preservation in a process that produces lactic acid found in such sour foods as pickled cucumbers, kimchi, and yogurt, as well as for producing alcoholic beverages such as wine and beer. Fermentation occurs within the gastrointestinal tracts of all animals, including humans.

Fermentation in progress: Bubbles of CO_2 form a froth on top of the fermentation mixture.

Below are some definitions of fermentation. They range from informal, general usages to more scientific definitions:

- Preservation methods for food via microorganisms (general use).

- Any process that produces alcoholic beverages or acidic dairy products (general use).

- Any large-scale microbial process occurring with or without air (common definition used in industry).

- Any energy-releasing metabolic process that takes place only under anaerobic conditions (becoming more scientific).

- Any metabolic process that releases energy from a sugar or other organic molecule, does not require oxygen or an electron transport system, and uses an organic molecule as the final electron acceptor (most scientific).

Biological Role

Along with photosynthesis and aerobic respiration, fermentation is a way of extracting energy from molecules, but it is the only one common to all bacteria and eukaryotes. It is therefore considered the oldest metabolic pathway, suitable for an environment that does not yet have oxygen.

Yeast, a form of fungus, occurs in almost any environment capable of supporting microbes, from the skins of fruits to the guts of insects and mammals and the deep ocean, and they harvest sugar-rich materials to produce ethanol and carbon dioxide.

The basic mechanism for fermentation remains present in all cells of higher organisms. Mammalian muscle carries out the fermentation that occurs during periods of intense exercise where oxygen supply becomes limited, resulting in the creation of lactic acid. In invertebrates, fermentation also produces succinate and alanine.

Fermentative bacteria play an essential role in the production of methane in habitats ranging from the rumens of cattle to sewage digesters and freshwater sediments. They produce hydrogen, carbon dioxide, formate and acetate and carboxylic acids; and then consortia of microbes convert the carbon dioxide and acetate to methane. Acetogenic bacteria oxidize the acids, obtaining more acetate and either hydrogen or formate. Finally, methanogens (which are in the domain *Archea*) convert acetate to methane.

Biochemical Overview

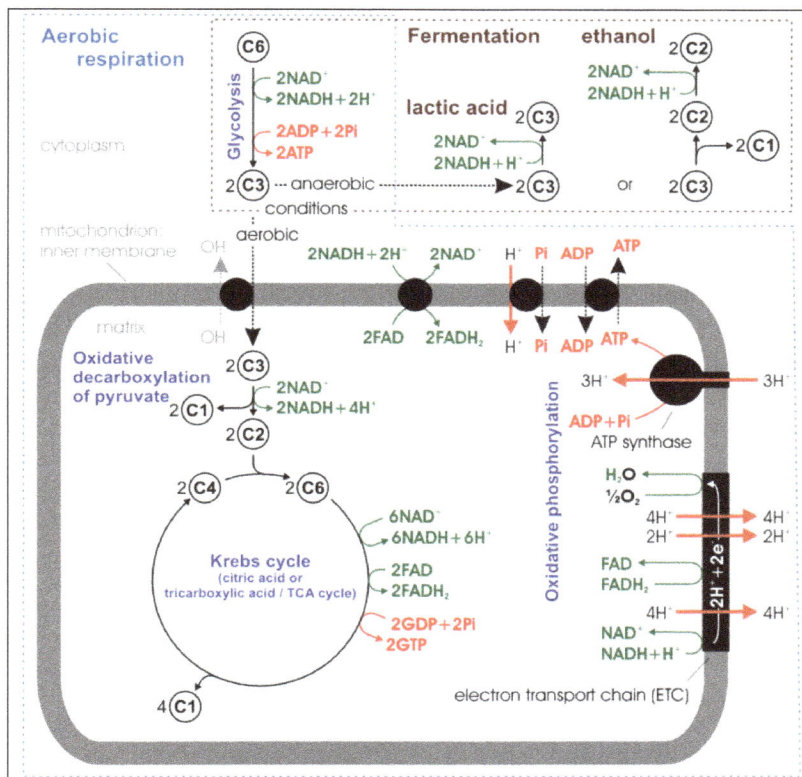

Comparison of aerobic respiration and most known fermentation types in eucaryotic cell. Numbers in circles indicate counts of carbon atoms in molecules, C6 is glucose $C_6H_{12}O_6$, C1 carbon dioxide CO_2. Mitochondrial outer membrane is omitted.

Fermentation reacts NADH with an endogenous, organic electron acceptor. Usually this is pyruvate formed from sugar through glycolysis. The reaction produces NAD+ and an organic product, typical examples being ethanol, lactic acid, carbon dioxide, and hydrogen gas (H_2). However, more exotic compounds can be produced by fermentation, such as butyric acid and acetone. Fermentation products contain chemical energy (they are not fully oxidized), but are considered waste products, since they cannot be metabolized further without the use of oxygen.

Although yeast carries out the fermentation in the production of ethanol in beers, wines, and other alcoholic drinks, this is not the only possible agent: bacteria carry out the fermentation in the production of xanthan gum.

Human Genome Project

The Human Genome Project (HGP) was an international scientific research project with the goal of determining the sequence of nucleotide base pairs that make up human DNA, and of identifying and mapping all of the genes of the human genome from both a physical and a functional standpoint. It remains the world's largest collaborative biological project. After the idea was picked up in 1984 by the US government when the planning started, the project formally launched in 1990 and was declared complete on April 14, 2003. Funding came from the US government through the National Institutes of Health (NIH) as well as numerous other groups from around the world. A parallel project was conducted outside government by the Celera Corporation, or Celera Genomics, which was formally launched in 1998. Most of the government-sponsored sequencing was performed in twenty universities and research centers in the United States, the United Kingdom, Japan, France, Germany and China.

The Human Genome Project originally aimed to map the nucleotides contained in a human haploid reference genome (more than three billion). The "genome" of any given individual is unique; mapping the "human genome" involved sequencing a small number of individuals and then assembling these together to get a complete sequence for each chromosome. Therefore, the finished human genome is a mosaic, not representing any one individual.

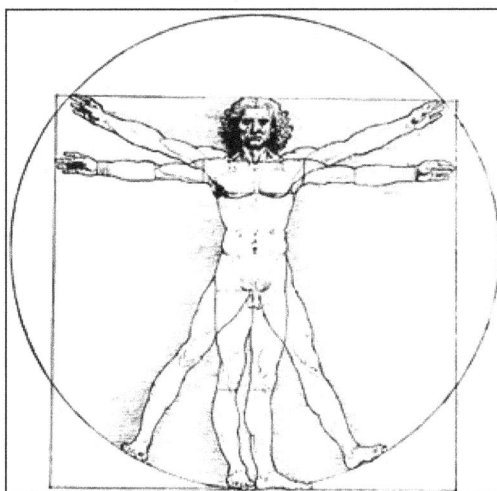

Logo of the HGP – the Vitruvian Man by Leonardo da Vinci.

The Human Genome Project was a 13-year-long, publicly funded project initiated in 1990 with the objective of determining the DNA sequence of the entire euchromatic human genome within 15 years. In May 1985, Robert Sinsheimer organized a workshop to discuss sequencing the human genome, but for a number of reasons the NIH was uninterested in pursuing the proposal. The following March, the Santa Fe Workshop was organized by Charles DeLisi and David Smith of the Department of Energy's Office of Health and Environmental Research (OHER). At the same time

Renato Dulbecco proposed whole genome sequencing in an essay in Science. James Watson followed two months later with a workshop held at the Cold Spring Harbor Laboratory.

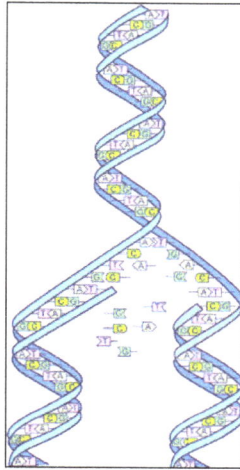

The fact that the Santa Fe workshop was motivated and supported by a Federal Agency opened a path, albeit a difficult and tortuous one, for converting the idea into a public policy in the United States. In a memo to the Assistant Secretary for Energy Research (Alvin Trivelpiece), Charles DeLisi, who was then Director of the OHER, outlined a broad plan for the project. This started a long and complex chain of events which led to approved reprogramming of funds that enabled the OHER to launch the Project in 1986, and to recommend the first line item for the HGP, which was in President Reagan's 1988 budget submission, and ultimately approved by the Congress. Of particular importance in Congressional approval was the advocacy of Senator Peter Domenici, whom DeLisi had befriended. Domenici chaired the Senate Committee on Energy and Natural Resources, as well as the Budget Committee, both of which were key in the DOE budget process. Congress added a comparable amount to the NIH budget, thereby beginning official funding by both agencies.

Alvin Trivelpiece sought and obtained the approval of DeLisi's proposal by Deputy Secretary William Flynn Martin. This chart was used in the spring of 1986 by Trivelpiece, then Director of the Office of Energy Research in the Department of Energy, to brief Martin and Under Secretary Joseph Salgado regarding his intention to reprogram $4 million to initiate the project with the approval of Secretary Herrington. This reprogramming was followed by a line item budget of $16 million in the Reagan Administration's 1987 budget submission to Congress. It subsequently passed both Houses. The Project was planned for 15 years.

Candidate technologies were already being considered for the proposed undertaking at least as early as 1979; Ronald W. Davis and colleagues of Stanford University submitted a proposal to NIH that year and it was turned down as being too ambitious.

In 1990, the two major funding agencies, DOE and NIH, developed a memorandum of understanding in order to coordinate plans and set the clock for the initiation of the Project to 1990. At that time, David Galas was Director of the renamed "Office of Biological and Environmental Research" in the U.S. Department of Energy's Office of Science and James Watson headed the NIH Genome Program. In 1993, Aristides Patrinos succeeded Galas and Francis Collins succeeded

James Watson, assuming the role of overall Project Head as Director of the U.S. National Institutes of Health (NIH) National Center for Human Genome Research (which would later become the National Human Genome Research Institute). A working draft of the genome was announced in 2000 and the papers describing it were published in February 2001. A more complete draft was published in 2003, and genome "finishing" work continued for more than a decade.

The $3-billion project was formally founded in 1990 by the US Department of Energy and the National Institutes of Health, and was expected to take 15 years. In addition to the United States, the international consortium comprised geneticists in the United Kingdom, France, Australia, China and myriad other spontaneous relationships. Adjusted for inflation the project cost roughly $5 billion.

Due to widespread international cooperation and advances in the field of genomics (especially in sequence analysis), as well as major advances in computing technology, a 'rough draft' of the genome was finished in 2000 (announced jointly by U.S. President Bill Clinton and the British Prime Minister Tony Blair on June 26, 2000). This first available rough draft assembly of the genome was completed by the Genome Bioinformatics Group at the University of California, Santa Cruz, primarily led by then graduate student Jim Kent. Ongoing sequencing led to the announcement of the essentially complete genome on April 14, 2003, two years earlier than planned. In May 2006, another milestone was passed on the way to completion of the project, when the sequence of the last chromosome was published in nature.

State of Completion

The project was not able to sequence all the DNA found in human cells. It sequenced only *euchromatic* regions of the genome, which make up 92.1% of the human genome. The other regions, called *heterochromatic*, are found in centromeres and telomeres, and were not sequenced under the project.

The Human Genome Project (HGP) was declared complete in April 2003. An initial rough draft of the human genome was available in June 2000 and by February 2001 a working draft had been completed and published followed by the final sequencing mapping of the human genome on April 14, 2003. Although this was reported to cover 99% of the euchromatic human genome with 99.99% accuracy, a major quality assessment of the human genome sequence was published on May 27, 2004 indicating over 92% of sampling exceeded 99.99% accuracy which was within the intended goal. Further analyses and papers on the HGP continue to occur.

In March 2009, the Genome Reference Consortium (GRC) released a more accurate version of the human genome, but that still left more than 300 gaps.

As of June 2019, the GRC still indicates 89 "unresolved" gaps, most of which are annotated as "stalled" or "under investigation/review".

Applications and Proposed Benefits

The sequencing of the human genome holds benefits for many fields, from molecular medicine to human evolution. The Human Genome Project, through its sequencing of the DNA, can help us understand diseases including: genotyping of specific viruses to direct appropriate treatment;

identification of mutations linked to different forms of cancer; the design of medication and more accurate prediction of their effects; advancement in forensic applied sciences; biofuels and other energy applications; agriculture, animal husbandry, bioprocessing; risk assessment; bioarcheology, anthropology and evolution. Another proposed benefit is the commercial development of genomics research related to DNA based products, a multibillion-dollar industry.

The sequence of the DNA is stored in databases available to anyone on the Internet. The U.S. National Center for Biotechnology Information (and sister organizations in Europe and Japan) house the gene sequence in a database known as GenBank, along with sequences of known and hypothetical genes and proteins. Other organizations, such as the UCSC Genome Browser at the University of California, Santa Cruz, and Ensembl present additional data and annotation and powerful tools for visualizing and searching it. Computer programs have been developed to analyze the data, because the data itself is difficult to interpret without such programs. Generally speaking, advances in genome sequencing technology have followed Moore's Law, a concept from computer science which states that integrated circuits can increase in complexity at an exponential rate. This means that the speeds at which whole genomes can be sequenced can increase at a similar rate, as was seen during the development of the above-mentioned Human Genome Project.

Techniques and Analysis

The process of identifying the boundaries between genes and other features in a raw DNA sequence is called genome annotation and is in the domain of bioinformatics. While expert biologists make the best annotators, their work proceeds slowly, and computer programs are increasingly used to meet the high-throughout demands of genome sequencing projects. Beginning in 2008, a new technology known as RNA-seq was introduced that allowed scientists to directly sequence the messenger RNA in cells. This replaced previous methods of annotation, which relied on inherent properties of the DNA sequence, with direct measurement, which was much more accurate. Today, annotation of the human genome and other genomes relies primarily on deep sequencing of the transcripts in every human tissue using RNA-seq. These experiments have revealed that over 90% of genes contain at least one and usually several alternative splice variants, in which the exons are combined in different ways to produce 2 or more gene products from the same locus.

The genome published by the HGP does not represent the sequence of every individual's genome. It is the combined mosaic of a small number of anonymous donors, all of European origin. The HGP genome is a scaffold for future work in identifying differences among individuals. Subsequent projects sequenced the genomes of multiple distinct ethnic groups, though as of today there is still only one "reference genome."

Findings

Key findings of the draft and complete genome sequences include:

- There are approximately 22,300 protein-coding genes in human beings, the same range as in other mammals.

- The human genome has significantly more segmental duplications (nearly identical, repeated sections of DNA) than had been previously suspected.

- At the time when the draft sequence was published fewer than 7% of protein families appeared to be vertebrate specific.

Accomplishment

The Human Genome Project was started in 1990 with the goal of sequencing and identifying all three billion chemical units in the human genetic instruction set, finding the genetic roots of disease and then developing treatments. It is considered a megaproject because the human genome has approximately 3.3 billion base pairs. With the sequence in hand, the next step was to identify the genetic variants that increase the risk for common diseases like cancer and diabetes.

The first printout of the human genome to be presented as a series of books, displayed at the Wellcome Collection, London.

It was far too expensive at that time to think of sequencing patients' whole genomes. So the National Institutes of Health embraced the idea for a "shortcut", which was to look just at sites on the genome where many people have a variant DNA unit. The theory behind the shortcut was that, since the major diseases are common, so too would be the genetic variants that caused them. Natural selection keeps the human genome free of variants that damage health before children are grown, the theory held, but fails against variants that strike later in life, allowing them to become quite common (In 2002 the National Institutes of Health started a $138 million project called the HapMap to catalog the common variants in European, East Asian and African genomes).

The genome was broken into smaller pieces; approximately 150,000 base pairs in length. These pieces were then ligated into a type of vector known as "bacterial artificial chromosomes", or BACs, which are derived from bacterial chromosomes which have been genetically engineered. The vectors containing the genes can be inserted into bacteria where they are copied by the bacterial DNA replication machinery. Each of these pieces was then sequenced separately as a small "shotgun" project and then assembled. The larger, 150,000 base pairs go together to create chromosomes. This is known as the "hierarchical shotgun" approach, because the genome is first broken into relatively large chunks, which are then mapped to chromosomes before being selected for sequencing.

Funding came from the US government through the National Institutes of Health in the United States, and a UK charity organization, the Wellcome Trust, as well as numerous other groups from around the world. The funding supported a number of large sequencing centers including those at

Whitehead Institute, the Wellcome Sanger Institute (then called The Sanger Centre) based at the Wellcome Genome Campus, Washington University in St. Louis, and Baylor College of Medicine.

The United Nations Educational, Scientific and Cultural Organization (UNESCO) served as an important channel for the involvement of developing countries in the Human Genome Project.

Public versus Private Approaches

In 1998, a similar, privately funded quest was launched by the American researcher Craig Venter, and his firm Celera Genomics. Venter was a scientist at the NIH during the early 1990s when the project was initiated. The $300,000,000 Celera effort was intended to proceed at a faster pace and at a fraction of the cost of the roughly $3 billion publicly funded project. The Celera approach was able to proceed at a much more rapid rate, and at a lower cost than the public project because it relied upon data made available by the publicly funded project.

Celera used a technique called whole genome shotgun sequencing, employing pairwise end sequencing, which had been used to sequence bacterial genomes of up to six million base pairs in length, but not for anything nearly as large as the three billion base pair human genome.

Celera initially announced that it would seek patent protection on "only 200–300" genes, but later amended this to seeking "intellectual property protection" on "fully-characterized important structures" amounting to 100–300 targets. The firm eventually filed preliminary ("place-holder") patent applications on 6,500 whole or partial genes. Celera also promised to publish their findings in accordance with the terms of the 1996 "Bermuda Statement", by releasing new data annually (the HGP released its new data daily), although, unlike the publicly funded project, they would not permit free redistribution or scientific use of the data. The publicly funded competitors were compelled to release the first draft of the human genome before Celera for this reason. On July 7, 2000, the UCSC Genome Bioinformatics Group released a first working draft on the web. The scientific community downloaded about 500 GB of information from the UCSC genome server in the first 24 hours of free and unrestricted access.

In March 2000, President Clinton announced that the genome sequence could not be patented, and should be made freely available to all researchers. The statement sent Celera's stock plummeting and dragged down the biotechnology-heavy Nasdaq. The biotechnology sector lost about $50 billion in market capitalization in two days.

Genome Donors

In the IHGSC international public-sector HGP, researchers collected blood (female) or sperm (male) samples from a large number of donors. Only a few of many collected samples were processed as DNA resources. Thus the donor identities were protected so neither donors nor scientists could know whose DNA was sequenced. DNA clones from many different libraries were used in the overall project, with most of those libraries being created by Pieter J. de Jong's. Much of the sequence (>70%) of the reference genome produced by the public HGP came from a single anonymous male donor from Buffalo, New York (code name RP11).

HGP scientists used white blood cells from the blood of two male and two female donors (randomly selected from 20 of each) – each donor yielding a separate DNA library. One of these libraries

(RP11) was used considerably more than others, due to quality considerations. One minor technical issue is that male samples contain just over half as much DNA from the sex chromosomes (one X chromosome and one Y chromosome) compared to female samples (which contain two X chromosomes). The other 22 chromosomes (the autosomes) are the same for both sexes.

Although the main sequencing phase of the HGP has been completed, studies of DNA variation continued in the International HapMap Project, whose goal was to identify patterns of single-nucleotide polymorphism (SNP) groups (called haplotypes, or "haps"). The DNA samples for the HapMap came from a total of 270 individuals: Yoruba people in Ibadan, Nigeria; Japanese people in Tokyo; Han Chinese in Beijing; and the French Centre d'Etude du Polymorphisme Humain (CEPH) resource, which consisted of residents of the United States having ancestry from Western and Northern Europe.

In the Celera Genomics private-sector project, DNA from five different individuals were used for sequencing. The lead scientist of Celera Genomics at that time, Craig Venter, later acknowledged that his DNA was one of 21 samples in the pool, five of which were selected for use.

In 2007, a team led by Jonathan Rothberg published James Watson's entire genome, unveiling the six-billion-nucleotide genome of a single individual for the first time.

Developments

The work on interpretation and analysis of genome data is still in its initial stages. It is anticipated that detailed knowledge of the human genome will provide new avenues for advances in medicine and biotechnology. Clear practical results of the project emerged even before the work was finished. For example, a number of companies, such as Myriad Genetics, started offering easy ways to administer genetic tests that can show predisposition to a variety of illnesses, including breast cancer, hemostasis disorders, cystic fibrosis, liver diseases and many others. Also, the etiologies for cancers, Alzheimer's disease and other areas of clinical interest are considered likely to benefit from genome information and possibly may lead in the long term to significant advances in their management.

There are also many tangible benefits for biologists. For example, a researcher investigating a certain form of cancer may have narrowed down their search to a particular gene. By visiting the human genome database on the World Wide Web, this researcher can examine what other scientists have written about this gene, including (potentially) the three-dimensional structure of its product, its function(s), its evolutionary relationships to other human genes, or to genes in mice or yeast or fruit flies, possible detrimental mutations, interactions with other genes, body tissues in which this gene is activated, and diseases associated with this gene or other datatypes. Further, deeper understanding of the disease processes at the level of molecular biology may determine new therapeutic procedures. Given the established importance of DNA in molecular biology and its central role in determining the fundamental operation of cellular processes, it is likely that expanded knowledge in this area will facilitate medical advances in numerous areas of clinical interest that may not have been possible without them.

The analysis of similarities between DNA sequences from different organisms is also opening new avenues in the study of evolution. In many cases, evolutionary questions can now be framed in terms of molecular biology; indeed, many major evolutionary milestones (the emergence of the ribosome and organelles, the development of embryos with body plans, the vertebrate immune system) can be related to the molecular level. Many questions about the similarities and differences between humans and our closest relatives (the primates, and indeed the other mammals) are expected to be illuminated by the data in this project.

The project inspired and paved the way for genomic work in other fields, such as agriculture. For example, by studying the genetic composition of Tritium aestivum, the world's most commonly used bread wheat, great insight has been gained into the ways that domestication has impacted the evolution of the plant. Which loci are most susceptible to manipulation, and how does this play out in evolutionary terms? Genetic sequencing has allowed these questions to be addressed for the first time, as specific loci can be compared in wild and domesticated strains of the plant. This will allow for advances in genetic modification in the future which could yield healthier and disease-resistant wheat crops, etc.

Ethical, Legal and Social Issues

At the onset of the Human Genome Project several ethical, legal, and social concerns were raised in regards to how increased knowledge of the human genome could be used to discriminate against people. One of the main concerns of most individuals was the fear that both employers and health insurance companies would refuse to hire individuals or refuse to provide insurance to people because of a health concern indicated by someone's genes. In 1996 the United States passed the Health Insurance Portability and Accountability Act (HIPAA) which protects against the unauthorized and non-consensual release of individually identifiable health information to any entity not actively engaged in the provision of healthcare services to a patient. Other nations passed no such protections.

Along with identifying all of the approximately 20,000–25,000 genes in the human genome, the Human Genome Project also sought to address the ethical, legal, and social issues that were created by the onset of the project. For that the Ethical, Legal, and Social Implications (ELSI) program was founded in 1990. Five percent of the annual budget was allocated to address the ELSI arising from the project. This budget started at approximately $1.57 million in the year 1990, but increased to approximately $18 million in the year 2014.

Whilst the project may offer significant benefits to medicine and scientific research, some authors have emphasized the need to address the potential social consequences of mapping the human genome. "Molecularising disease and their possible cure will have a profound impact on what patients expect from medical help and the new generation of doctors' perception of illness."

References

- Cell-culture: microscopemaster.com, Retrieved 13 March, 2019

- Zumdahl, steven s. Zumdahl, susan a. (2008). Chemistry (8th ed.). Ca: cengage learning. Isbn 978-0547125329

- Tissue-culture: microscopemaster.com, Retrieved 29 June, 2019

- Pratt, donald voet, judith g. Voet, charlotte w. (2013). Fundamentals of biochemistry : life at the molecular level (4th ed.). Hoboken, nj: wiley. Isbn 978-0470547847

- Cloning, science: britannica.com, Retrieved 2 July, 2019

- Buffum, burt c. (2008). Arid agriculture; a hand-book for the western farmer and stockman. Read books. P. 232. Isbn 978-1-4086-6710-1

- Jain, h. K.; kharkwal, m. C. (2004). Plant breeding - mendelian to molecular approaches. Boston, london, dordecht: kluwer academic publishers. Isbn 978-1-4020-1981-4

4
Applications of Biotechnology

Biotechnology is applied in various disciplines such as genetic engineering, biomimetics and biomanufacturing. It also plays an important role in the functioning of a biobased economy. The diverse applications of biotechnology in these areas have been thoroughly discussed in this chapter.

Genetic Engineering

Genetic engineering is the artificial manipulation, modification, and recombination of DNA or other nucleic acid molecules in order to modify an organism or population of organisms.

The term genetic engineering initially referred to various techniques used for the modification or manipulation of organisms through the processes of heredity and reproduction. As such, the term embraced both artificial selection and all the interventions of biomedical techniques, among them artificial insemination, in vitro fertilization (e.g., "test-tube" babies), cloning, and gene manipulation. In the latter part of the 20th century, however, the term came to refer more specifically to methods of recombinant DNA technology (or gene cloning), in which DNA molecules from two or more sources are combined either within cells or in vitro and are then inserted into host organisms in which they are able to propagate.

The possibility for recombinant DNA technology emerged with the discovery of restriction enzymes in 1968 by Swiss microbiologist Werner Arber. The following year American microbiologist Hamilton O. Smith purified so-called type II restriction enzymes, which were found to be essential to genetic engineering for their ability to cleave a specific site within the DNA (as opposed to type I restriction enzymes, which cleave DNA at random sites). Drawing on Smith's work, American molecular biologist Daniel Nathans helped advance the technique of DNA recombination in 1970–71 and demonstrated that type II enzymes could be useful in genetic studies. Genetic engineering based on recombination was pioneered in 1973 by American biochemists Stanley N. Cohen and Herbert W. Boyer, who were among the first to cut DNA into fragments, rejoin different fragments, and insert the new genes into E. coli bacteria, which then reproduced.

Process and Techniques

Most recombinant DNA technology involves the insertion of foreign genes into the plasmids of common laboratory strains of bacteria. Plasmids are small rings of DNA; they are not part of the bacterium's chromosome (the main repository of the organism's genetic information). Nonetheless, they are capable of directing protein synthesis, and, like chromosomal DNA, they are reproduced and passed on to the bacterium's progeny. Thus, by incorporating foreign DNA (for example, a mammalian gene) into a bacterium, researchers can obtain an almost limitless number of copies of the inserted gene. Furthermore, if the inserted gene is operative (i.e., if it directs protein synthesis), the modified bacterium will produce the protein specified by the foreign DNA.

A subsequent generation of genetic engineering techniques that emerged in the early 21st century centred on gene editing. Gene editing, based on a technology known as CRISPR-Cas9, allows researchers to customize a living organism's genetic sequence by making very specific changes to its DNA. Gene editing has a wide array of applications, being used for the genetic modification of crop plants and livestock and of laboratory model organisms (e.g., mice). The correction of genetic errors associated with disease in animals suggests that gene editing has potential applications in gene therapy for humans.

Applications

Genetic engineering has advanced the understanding of many theoretical and practical aspects of gene function and organization. Through recombinant DNA techniques, bacteria have been created that are capable of synthesizing human insulin, human growth hormone, alpha interferon, a hepatitis B vaccine, and other medically useful substances. Plants may be genetically adjusted to enable them to fix nitrogen, and genetic diseases can possibly be corrected by replacing dysfunctional genes with normally functioning genes. Nevertheless, special concern has been focused on such achievements for fear that they might result in the introduction of unfavourable and possibly dangerous traits into microorganisms that were previously free of them—e.g., resistance to antibiotics, production of toxins, or a tendency to cause disease. Likewise, the application of gene editing in humans has raised ethical concerns, particularly regarding its potential use to alter traits such as intelligence and beauty.

Controversy

In 1980 the "new" microorganisms created by recombinant DNA research were deemed patentable, and in 1986 the U.S. Department of Agriculture approved the sale of the first living genetically altered organism—a virus, used as a pseudorabies vaccine, from which a single gene had been cut. Since then several hundred patents have been awarded for genetically altered bacteria and plants. Patents on genetically engineered and genetically modified organisms, particularly crops and other foods, however, were a contentious issue, and they remained so into the first part of the 21st century.

Genetically Modified Crops

Genetically modified crops (GM crops) are plants used in agriculture, the DNA of which has been modified using genetic engineering methods. In most cases, the aim is to introduce a new trait to the plant which does not occur naturally in the species. Examples in food crops include resistance to certain pests, diseases, environmental conditions, reduction of spoilage, resistance to chemical treatments (e.g. resistance to a herbicide), or improving the nutrient profile of the crop. Examples in non-food crops include production of pharmaceutical agents, biofuels, and other industrially useful goods, as well as for bioremediation.

Farmers have widely adopted GM technology. Acreage increased from 1.7 million hectares in 1996 to 185.1 million hectares in 2016, some 12% of global cropland. As of 2016, major crop (soybean, maize, canola and cotton) traits consist of herbicide tolerance (95.9 million hectares) insect resistance (25.2 million hectares), or both (58.5 million hectares). In 2015, 53.6 million ha of GM maize were under cultivation (almost 1/3 of the maize crop). GM maize outperformed its predecessors: yield was 5.6 to 24.5% higher with less mycotoxins (−28.8%), fumonisin (−30.6%) and thricotecens (−36.5%). Non-target organisms were unaffected, except for *Braconidae*, represented by a parasitoid of European corn borer, the target of *Lepidoptera* active Bt maize. Biogeochemical parameters such as lignin content did not vary, while biomass decomposition was higher.

A 2014 meta-analysis concluded that GM technology adoption had reduced chemical pesticide use by 37%, increased crop yields by 22%, and increased farmer profits by 68%. This reduction in pesticide use has been ecologically beneficial, but benefits may be reduced by overuse. Yield gains and pesticide reductions are larger for insect-resistant crops than for herbicide-tolerant crops. Yield and profit gains are higher in developing countries than in developed countries.

There is a scientific consensus that currently available food derived from GM crops poses no greater risk to human health than conventional food, but that each GM food needs to be tested on a case-by-case basis before introduction. Nonetheless, members of the public are much less likely than scientists to perceive GM foods as safe. The legal and regulatory status of GM foods varies by country, with some nations banning or restricting them, and others permitting them with widely differing degrees of regulation.

However, opponents have objected to GM crops on grounds including environmental impacts, food safety, whether GM crops are needed to address food needs, whether they are sufficiently accessible to farmers in developing countries and concerns over subjecting crops to intellectual property law. Safety concerns led 38 countries, including 19 in Europe, to officially prohibit their cultivation.

Methods

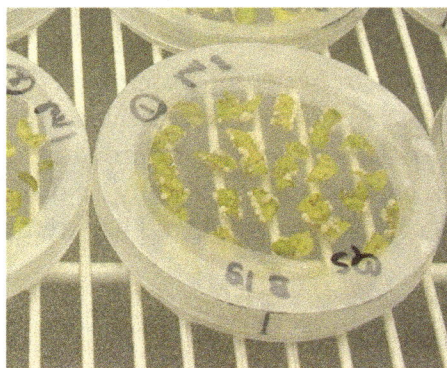

Plants (Solanum chacoense) being transformed using agrobacterium.

Genetically engineered crops have genes added or removed using genetic engineering techniques, originally including gene guns, electroporation, microinjection and agrobacterium. More recently, CRISPR and TALEN offered much more precise and convenient editing techniques.

Gene guns (also known as biolistics) "shoot" (direct high energy particles or radiations against) target genes into plant cells. It is the most common method. DNA is bound to tiny particles of gold or tungsten which are subsequently shot into plant tissue or single plant cells under high pressure. The accelerated particles penetrate both the cell wall and membranes. The DNA separates from the metal and is integrated into plant DNA inside the nucleus. This method has been applied successfully for many cultivated crops, especially monocots like wheat or maize, for which transformation using *Agrobacterium tumefaciens* has been less successful. The major disadvantage of this procedure is that serious damage can be done to the cellular tissue.

Agrobacterium tumefaciens-mediated transformation is another common technique. Agrobacteria are natural plant parasites. Their natural ability to transfer genes provides another engineering method. To create a suitable environment for themselves, these Agrobacteria insert their genes into plant hosts, resulting in a proliferation of modified plant cells near the soil level (crown gall). The genetic information for tumor growth is encoded on a mobile, circular DNA fragment (plasmid). When *Agrobacterium* infects a plant, it transfers this T-DNA to a random site in the plant genome. When used in genetic engineering the bacterial T-DNA is removed from the bacterial plasmid and replaced with the desired foreign gene. The bacterium is a vector, enabling transportation of foreign genes into plants. This method works especially well for dicotyledonous plants like potatoes, tomatoes, and tobacco. Agrobacteria infection is less successful in crops like wheat and maize.

Electroporation is used when the plant tissue does not contain cell walls. In this technique, "DNA enters the plant cells through miniature pores which are temporarily caused by electric pulses."

Microinjection is used to directly inject foreign DNA into cells.

Plant scientists, backed by results of modern comprehensive profiling of crop composition, point out that crops modified using GM techniques are less likely to have unintended changes than are conventionally bred crops.

In research tobacco and *Arabidopsis thaliana* are the most frequently modified plants, due to well-developed transformation methods, easy propagation and well studied genomes. They serve as model organisms for other plant species.

Introducing new genes into plants requires a promoter specific to the area where the gene is to be expressed. For instance, to express a gene only in rice grains and not in leaves, an endosperm-specific promoter is used. The codons of the gene must be optimized for the organism due to codon usage bias.

Types of Modifications

Transgenic

Transgenic maize containing a gene from the bacteria Bacillus thuringiensis.

Transgenic plants have genes inserted into them that are derived from another species. The inserted genes can come from species within the same kingdom (plant to plant), or between kingdoms (for example, bacteria to plant). In many cases the inserted DNA has to be modified slightly in order to be correctly and efficiently expressed in the host organism. Transgenic plants are used to express proteins, like the cry toxins from *B. thuringiensis*, herbicide-resistant genes, antibodies, and antigens for vaccinations. A study led by the European Food Safety Authority (EFSA) also found viral genes in transgenic plants.

Transgenic carrots have been used to produce the drug Taliglucerase alfa which is used to treat Gaucher's disease. In the laboratory, transgenic plants have been modified to increase photosynthesis (currently about 2% at most plants versus the theoretic potential of 9–10%). This is possible by changing the rubisco enzyme (i.e. changing C_3 plants into C_4 plants), by placing the rubisco in a carboxysome, by adding CO_2 pumps in the cell wall, or by changing the leaf form or size. Plants have been engineered to exhibit bioluminescence that may become a sustainable alternative to electric lighting.

Cisgenic

Cisgenic plants are made using genes found within the same species or a closely related one, where conventional plant breeding can occur. Some breeders and scientists argue that cisgenic modification is useful for plants that are difficult to crossbreed by conventional means (such as potatoes), and that plants in the cisgenic category should not require the same regulatory scrutiny as transgenics.

Subgenic

Genetically modified plants can also be developed using gene knockdown or gene knockout to alter the genetic makeup of a plant without incorporating genes from other plants. In 2014, Chinese researcher Gao Caixia filed patents on the creation of a strain of wheat that is resistant to powdery mildew. The strain lacks genes that encode proteins that repress defenses against the mildew. The researchers deleted all three copies of the genes from wheat's hexaploid genome. Gao used the TALENs and CRISPR gene editing tools without adding or changing any other genes. No field trials were immediately planned. The CRISPR technique has also been used by Penn State researcher Yinong Yang to modify white button mushrooms (*Agaricus bisporus*) to be non-browning, and by DuPont Pioneer to make a new variety of corn.

Multiple Trait Integration

With multiple trait integration, several new traits may be integrated into a new crop.

Economics

GM food's economic value to farmers is one of its major benefits, including in developing nations. A 2010 study found that Bt corn provided economic benefits of $6.9 billion over the previous 14 years in five Midwestern states. The majority ($4.3 billion) accrued to farmers producing non-Bt corn. This was attributed to European corn borer populations reduced by exposure to Bt corn, leaving fewer to attack conventional corn nearby. Agriculture economists calculated that "world surplus [increased by] $240.3 million for 1996. Of this total, the largest share (59%) went to U.S. farmers. Seed company Monsanto received the next largest share (21%), followed by US consumers (9%), the rest of the world (6%), and the germplasm supplier, Delta & Pine Land Company of Mississippi (5%)."

According to the International Service for the Acquisition of Agri-biotech Applications (ISAAA), in 2014 approximately 18 million farmers grew biotech crops in 28 countries; about 94% of the farmers were resource-poor in developing countries. 53% of the global biotech crop area of 181.5 million hectares was grown in 20 developing countries. PG Economics comprehensive 2012 study concluded that GM crops increased farm incomes worldwide by $14 billion in 2010, with over half this total going to farmers in developing countries.

Critics challenged the claimed benefits to farmers over the prevalence of biased observers and by the absence of randomized controlled trials. The main Bt crop grown by small farmers in developing countries is cotton. A 2006 review of Bt cotton findings by agricultural economists concluded, "the overall balance sheet, though promising, is mixed. Economic returns are highly variable over years, farm type, and geographical location".

In 2013 the European Academies Science Advisory Council (EASAC) asked the EU to allow the development of agricultural GM technologies to enable more sustainable agriculture, by employing fewer land, water, and nutrient resources. EASAC also criticizes the EU's "time-consuming and expensive regulatory framework" and said that the EU had fallen behind in the adoption of GM technologies.

Participants in agriculture business markets include seed companies, agrochemical companies, distributors, farmers, grain elevators and universities that develop new crops/traits and whose agricultural extensions advise farmers on best practices. According to a 2012 review based on data from the late 1990s and early 2000s, much of the GM crop grown each year is used for livestock feed and increased demand for meat leads to increased demand for GM feed crops. Feed grain usage as a percentage of total crop production is 70% for corn and more than 90% of oil seed meals such as soybeans. About 65 million metric tons of GM corn grains and about 70 million metric tons of soybean meals derived from GM soybean become feed.

In 2014 the global value of biotech seed was US$15.7 billion; US$11.3 billion (72%) was in industrial countries and US$4.4 billion (28%) was in the developing countries. In 2009, Monsanto had $7.3 billion in sales of seeds and from licensing its technology; DuPont, through its Pioneer subsidiary, was the next biggest company in that market. As of 2009, the overall Roundup line of products including the GM seeds represented about 50% of Monsanto's business.

Some patents on GM traits have expired, allowing the legal development of generic strains that include these traits. For example, generic glyphosate-tolerant GM soybean is now available. Another impact is that traits developed by one vendor can be added to another vendor's proprietary strains, potentially increasing product choice and competition. The patent on the first type of *Roundup Ready* crop that Monsanto produced (soybeans) expired in 2014 and the first harvest of off-patent soybeans occurs in the spring of 2015. Monsanto has broadly licensed the patent to other seed companies that include the glyphosate resistance trait in their seed products. About 150 companies have licensed the technology, including Syngenta and DuPont Pioneer.

Yield

In 2014, the largest review yet concluded that GM crops' effects on farming were positive. The meta-analysis considered all published English-language examinations of the agronomic and economic impacts between 1995 and March 2014 for three major GM crops: soybean, maize, and cotton. The study found that herbicide-tolerant crops have lower production costs, while for insect-resistant crops the reduced pesticide use was offset by higher seed prices, leaving overall production costs about the same.

Yields increased 9% for herbicide tolerance and 25% for insect resistant varieties. Farmers who adopted GM crops made 69% higher profits than those who did not. The review found that GM crops help farmers in developing countries, increasing yields by 14 percentage points.

The researchers considered some studies that were not peer-reviewed and a few that did not report sample sizes. They attempted to correct for publication bias, by considering sources beyond academic journals. The large data set allowed the study to control for potentially confounding variables such as fertilizer use. Separately, they concluded that the funding source did not influence study results.

Traits

GM crops grown today, or under development, have been modified with various traits. These traits include improved shelf life, disease resistance, stress resistance, herbicide resistance, pest resistance, production of useful goods such as biofuel or drugs, and ability to absorb toxins and for use in bioremediation of pollution.

Recently, research and development has been targeted to enhancement of crops that are locally important in developing countries, such as insect-resistant cowpea for Africa and insect-resistant brinjal (eggplant).

Extended Shelf Life

The first genetically modified crop approved for sale in the U.S. was the *FlavrSavr* tomato, which had a longer shelf life. First sold in 1994, FlavrSavr tomato production ceased in 1997. It is no longer on the market.

In November 2014, the USDA approved a GM potato that prevents bruising.

In February 2015 Arctic Apples were approved by the USDA, becoming the first genetically modified apple approved for US sale. Gene silencing was used to reduce the expression of polyphenol oxidase (PPO), thus preventing enzymatic browning of the fruit after it has been sliced open. The trait was added to Granny Smith and Golden Delicious varieties. The trait includes a bacterial antibiotic resistance gene that provides resistance to the antibiotic kanamycin. The genetic engineering involved cultivation in the presence of kanamycin, which allowed only resistant cultivars to survive.

Improved Photosynthesis

Plants use non-photochemical quenching (NPQ) to protect them from excessive amounts of sunlight. Plants can switch on the quenching mechanism almost instantaneously but it takes much longer for it to switch off again. During the time that it is switched off, the amount of energy that is wasted increases. A genetic modification in three genes allows to correct this (in a trial with tobacco plants). As a result, yields were 14-20% higher, in terms of the weight of the dry leaves harvested. The plants had bigger leaves, were taller and had more vigorous roots.

Another improvement that can be made on the photosynthesis process (with C3 pathway plants) is on photorespiration. By inserting the C4 pathway into C3 plants, productivity may increase by as much as 50% for cereal crops, such as rice.

Improved Nutritional Value

Edible Oils

Some GM soybeans offer improved oil profiles for processing. Camelina sativa has been modified to produce plants that accumulate high levels of oils similar to fish oils.

Vitamin Enrichment

Golden rice, developed by the International Rice Research Institute (IRRI), provides greater amounts of vitamin A targeted at reducing vitamin A deficiency. As of January 2016, golden rice has not yet been grown commercially in any country.

Toxin Reduction

A genetically modified cassava under development offers lower cyanogen glucosides and enhanced protein and other nutrients (called BioCassava).

In November 2014, the USDA approved a potato that prevents bruising and produces less acrylamide when fried. They do not employ genes from non-potato species. The trait was added to the Russet Burbank, Ranger Russet and Atlantic varieties.

Stress Resistance

Plants have been engineered to tolerate non-biological stressors, such as drought, frost, and high soil salinity. In 2011, Monsanto's DroughtGard maize became the first drought-resistant GM crop to receive US marketing approval.

Drought resistance occurs by modifying the plant's genes responsible for the mechanism known as the crassulacean acid metabolism (CAM), which allows the plants to survive despite low water levels. This holds promise for water-heavy crops such as rice, wheat, soybeans and poplar to accelerate their adaptation to water-limited environments. Several salinity tolerance mechanisms have been identified in salt-tolerant crops. For example, rice, canola and tomato crops have been genetically modified to increase their tolerance to salt stress.

Herbicides

Glyphosate

As of 1999, the most prevalent GM trait was glyphosate-tolerance. Glyphosate (the active ingredient in Roundup and other herbicide products) kills plants by interfering with the shikimate pathway in plants, which is essential for the synthesis of the aromatic amino acids phenylalanine, tyrosine, and tryptophan. The shikimate pathway is not present in animals, which instead obtain aromatic amino acids from their diet. More specifically, glyphosate inhibits the enzyme 5-enolpyruvylshikimate-3-phosphate synthase (EPSPS).

This trait was developed because the herbicides used on grain and grass crops at the time were highly toxic and not effective against narrow-leaved weeds. Thus, developing crops that could withstand spraying with glyphosate would both reduce environmental and health risks, and give an agricultural edge to the farmer.

Some micro-organisms have a version of EPSPS that is resistant to glyphosate inhibition. One of these was isolated from an *Agrobacterium* strain CP4 (CP4 EPSPS) that was resistant to glyphosate. The CP4 EPSPS gene was engineered for plant expression by fusing the 5' end of the gene to a chloroplast transit peptide derived from the petunia EPSPS. This transit peptide was used because it had shown previously an ability to deliver bacterial EPSPS to the chloroplasts of other plants. This CP4 EPSPS gene was cloned and transfected into soybeans.

The plasmid used to move the gene into soybeans was PV-GMGTO4. It contained three bacterial genes, two CP4 EPSPS genes, and a gene encoding beta-glucuronidase (GUS) from *Escherichia coli* as a marker. The DNA was injected into the soybeans using the particle acceleration method. Soybean cultivar A5403 was used for the transformation.

Bromoxynil

Tobacco plants have been engineered to be resistant to the herbicide bromoxynil.

Glufosinate

Crops have been commercialized that are resistant to the herbicide glufosinate, as well. Crops engineered for resistance to multiple herbicides to allow farmers to use a mixed group of two, three, or four different chemicals are under development to combat growing herbicide resistance.

2,4-D

In October 2014 the US EPA registered Dow's Enlist Duo maize, which is genetically modified to be resistant to both glyphosate and 2,4-D, in six states. Inserting a bacterial aryloxyalkanoate dioxygenase gene, *aad1* makes the corn resistant to 2,4-D. The USDA had approved maize and soybeans with the mutation in September 2014.

Dicamba

Monsanto has requested approval for a stacked strain that is tolerant of both glyphosate and dicamba. The request includes plans for avoiding herbicide drift to other crops. Significant damage to other non-resistant crops occurred from dicamba formulations intended to reduce volatilization drifting when sprayed on resistant soybeans in 2017. The newer dicamba formulation labels specify to not spray when average wind speeds are above 10–15 miles per hour (16–24 km/h) to avoid particle drift, average wind speeds below 3 miles per hour (4.8

km/h) to avoid temperature inversions, and rain or high temperatures are in the next day forecast. However, these conditions typically only occur during June and July for a few hours at a time.

Pest Resistance

Insects

Tobacco, corn, rice and some other crops have been engineered to express genes encoding for insecticidal proteins from Bacillus thuringiensis (Bt). The introduction of Bt crops during the period between 1996 and 2005 has been estimated to have reduced the total volume of insecticide active ingredient use in the United States by over 100 thousand tons. This represents a 19.4% reduction in insecticide use.

In the late 1990s, a genetically modified potato that was resistant to the Colorado potato beetle was withdrawn because major buyers rejected it, fearing consumer opposition.

Viruses

Papaya, potatoes, and squash have been engineered to resist viral pathogens such as cucumber mosaic virus which, despite its name, infects a wide variety of plants. Virus resistant papaya were developed in response to a papaya ringspot virus (PRV) outbreak in Hawaii in the late 1990s. They incorporate PRV DNA. By 2010, 80% of Hawaiian papaya plants were genetically modified.

Potatoes were engineered for resistance to potato leaf roll virus and Potato virus Y in 1998. Poor sales led to their market withdrawal after three years.

Yellow squash that were resistant to at first two, then three viruses were developed, beginning in the 1990s. The viruses are watermelon, cucumber and zucchini/courgette yellow mosaic. Squash was the second GM crop to be approved by US regulators. The trait was later added to zucchini.

Many strains of corn have been developed in recent years to combat the spread of Maize dwarf mosaic virus, a costly virus that causes stunted growth which is carried in Johnson grass and spread by aphid insect vectors. These strands are commercially available although the resistance is not standard among GM corn variants.

By-products

Drugs

In 2012, the FDA approved the first plant-produced pharmaceutical, a treatment for Gaucher's Disease. Tobacco plants have been modified to produce therapeutic antibodies.

Biofuel

Algae is under development for use in biofuels. Researchers in Singapore were working on GM jatropha for biofuel production. Syngenta has USDA approval to market a maize trademarked

Enogen that has been genetically modified to convert its starch to sugar for ethanol. Some trees have been genetically modified to either have less lignin, or to express lignin with chemically labile bonds. Lignin is the critical limiting factor when using wood to make bio-ethanol because lignin limits the accessibility of cellulose microfibrils to depolymerization by enzymes. Besides with trees, the chemically labile lignin bonds are also very useful for cereal crops such as maize, barley, and oats.

Materials

Companies and labs are working on plants that can be used to make bioplastics. Potatoes that produce industrially useful starches have been developed as well. Oilseed can be modified to produce fatty acids for detergents, substitute fuels and petrochemicals.

Bioremediation

Scientists at the University of York developed a weed (*Arabidopsis thaliana*) that contains genes from bacteria that could clean TNT and RDX-explosive soil contaminants in 2011. 16 million hectares in the US (1.5% of the total surface) are estimated to be contaminated with TNT and RDX. However *A. thaliana* was not tough enough for use on military test grounds. Modifications in 2016 included switchgrass and bentgrass.

Genetically modified plants have been used for bioremediation of contaminated soils. Mercury, selenium and organic pollutants such as polychlorinated biphenyls (PCBs).

Marine environments are especially vulnerable since pollution such as oil spills are not containable. In addition to anthropogenic pollution, millions of tons of petroleum annually enter the marine environment from natural seepages. Despite its toxicity, a considerable fraction of petroleum oil entering marine systems is eliminated by the hydrocarbon-degrading activities of microbial communities. Particularly successful is a recently discovered group of specialists, the so-called hydrocarbonoclastic bacteria (HCCB) that may offer useful genes.

Asexual Reproduction

Crops such as maize reproduce sexually each year. This randomizes which genes get propagated to the next generation, meaning that desirable traits can be lost. To maintain a high-quality crop, some farmers purchase seeds every year. Typically, the seed company maintains two inbred varieties and crosses them into a hybrid strain that is then sold. Related plants like sorghum and gamma grass are able to perform apomixis, a form of asexual reproduction that keeps the plant's DNA intact. This trait is apparently controlled by a single dominant gene, but traditional breeding has been unsuccessful in creating asexually-reproducing maize. Genetic engineering offers another route to this goal. Successful modification would allow farmers to replant harvested seeds that retain desirable traits, rather than relying on purchased seed.

Other

Genetic modifications to some crops also exist, which make it easier to process the crop. Also, some crops (such as tomatoes) have been genetic modified to contain no seed at all.

Development

The number of USDA-approved field releases for testing grew from 4 in 1985 to 1,194 in 2002 and averaged around 800 per year thereafter. The number of sites per release and the number of gene constructs (ways that the gene of interest is packaged together with other elements) – have rapidly increased since 2005. Releases with agronomic properties (such as drought resistance) jumped from 1,043 in 2005 to 5,190 in 2013. As of September 2013, about 7,800 releases had been approved for corn, more than 2,200 for soybeans, more than 1,100 for cotton, and about 900 for potatoes. Releases were approved for herbicide tolerance (6,772 releases), insect resistance (4,809), product quality such as flavor or nutrition (4,896), agronomic properties like drought resistance (5,190), and virus/fungal resistance (2,616). The institutions with the most authorized field releases include Monsanto with 6,782, Pioneer/DuPont with 1,405, Syngenta with 565, and USDA's Agricultural Research Service with 370. As of September 2013 USDA had received proposals for releasing GM rice, squash, plum, rose, tobacco, flax, and chicory.

Farming Practices

Resistance

1. Bacillus Thuringiensis:

Constant exposure to a toxin creates evolutionary pressure for pests resistant to that toxin. Over-reliance on glyphosate and a reduction in the diversity of weed management practices allowed the spread of glyphosate resistance in 14 weed species in the US, and in soybeans.

To reduce resistance to *Bacillus thuringiensis* (Bt) crops, the 1996 commercialization of transgenic cotton and maize came with a management strategy to prevent insects from becoming resistant. Insect resistance management plans are mandatory for Bt crops. The aim is to encourage a large population of pests so that any (recessive) resistance genes are diluted within the population. Resistance lowers evolutionary fitness in the absence of the stressor, Bt. In refuges, non-resistant strains outcompete resistant ones.

With sufficiently high levels of transgene expression, nearly all of the heterozygotes (S/s), i.e., the largest segment of the pest population carrying a resistance allele, will be killed before maturation, thus preventing transmission of the resistance gene to their progeny. Refuges (i. e., fields of non-transgenic plants) adjacent to transgenic fields increases the likelihood that homozygous resistant (s/s) individuals and any surviving heterozygotes will mate with susceptible (S/S) individuals from the refuge, instead of with other individuals carrying the resistance allele. As a result, the resistance gene frequency in the population remains lower.

Complicating factors can affect the success of the high-dose/refuge strategy. For example, if the temperature is not ideal, thermal stress can lower Bt toxin production and leave the plant more susceptible. More importantly, reduced late-season expression has been documented, possibly resulting from DNA methylation of the promoter. The success of the high-dose/refuge strategy has successfully maintained the value of Bt crops. This success has depended on factors independent of management strategy, including low initial resistance allele frequencies, fitness costs associated with resistance, and the abundance of non-Bt host plants outside the refuges.

Companies that produce Bt seed are introducing strains with multiple Bt proteins. Monsanto did this with Bt cotton in India, where the product was rapidly adopted. Monsanto has also; in an attempt to simplify the process of implementing refuges in fields to comply with Insect Resistance Management(IRM) policies and prevent irresponsible planting practices; begun marketing seed bags with a set proportion of refuge (non-transgenic) seeds mixed in with the Bt seeds being sold. Coined "Refuge-In-a-Bag" (RIB), this practice is intended to increase farmer compliance with refuge requirements and reduce additional labor needed at planting from having separate Bt and refuge seed bags on hand. This strategy is likely to reduce the likelihood of Bt-resistance occurring for corn rootworm, but may increase the risk of resistance for lepidopteran corn pests, such as European corn borer. Increased concerns for resistance with seed mixtures include partially resistant larvae on a Bt plant being able to move to a susceptible plant to survive or cross pollination of refuge pollen on to Bt plants that can lower the amount of Bt expressed in kernels for ear feeding insects.

2. Herbicide Resistance:

Best management practices (BMPs) to control weeds may help delay resistance. BMPs include applying multiple herbicides with different modes of action, rotating crops, planting weed-free seed, scouting fields routinely, cleaning equipment to reduce the transmission of weeds to other fields, and maintaining field borders. The most widely planted GM crops are designed to tolerate herbicides. By 2006 some weed populations had evolved to tolerate some of the same herbicides. Palmer amaranth is a weed that competes with cotton. A native of the southwestern US, it traveled east and was first found resistant to glyphosate in 2006, less than 10 years after GM cotton was introduced.

Plant Protection

Farmers generally use less insecticide when they plant Bt-resistant crops. Insecticide use on corn farms declined from 0.21 pound per planted acre in 1995 to 0.02 pound in 2010. This is consistent with the decline in European corn borer populations as a direct result of Bt corn and cotton. The establishment of minimum refuge requirements helped delay the evolution of Bt resistance. However, resistance appears to be developing to some Bt traits in some areas.

Tillage

By leaving at least 30% of crop residue on the soil surface from harvest through planting, conservation tillage reduces soil erosion from wind and water, increases water retention, and reduces soil degradation as well as water and chemical runoff. In addition, conservation tillage reduces the carbon footprint of agriculture. A 2014 review covering 12 states from 1996 to 2006, found that a 1% increase in herbicde-tolerant (HT) soybean adoption leads to a 0.21% increase in conservation tillage and a 0.3% decrease in quality-adjusted herbicide use.

Regulation

The regulation of genetic engineering concerns the approaches taken by governments to assess and manage the risks associated with the development and release of genetically modified crops. There are differences in the regulation of GM crops between countries, with some of the most marked differences occurring between the US and Europe. Regulation varies in a given country depending on the intended use of each product. For example, a crop not intended for food use is generally not reviewed by authorities responsible for food safety.

Production

In 2013, GM crops were planted in 27 countries; 19 were developing countries and 8 were developed countries. 2013 was the second year in which developing countries grew a majority (54%) of the total GM harvest. 18 million farmers grew GM crops; around 90% were small-holding farmers in developing countries.

Country	2013– GM planted area (million hectares)	Biotech crops
US	70.1	Maize, Soybean, Cotton, Canola, Sugarbeet, Alfalfa, Papaya, Squash
Brazil	40.3	Soybean, Maize, Cotton
Argentina	24.4	Soybean, Maize, Cotton
India	11.0	Cotton
Canada	10.8	Canola, Maize, Soybean, Sugarbeet
Total	175.2	----

The United States Department of Agriculture (USDA) reports every year on the total area of GM crop varieties planted in the United States. According to National Agricultural Statistics Service, the states published in these tables represent 81–86 percent of all corn planted area, 88–90 percent of all soybean planted area, and 81–93 percent of all upland cotton planted area (depending on the year).

Global estimates are produced by the International Service for the Acquisition of Agri-biotech Applications (ISAAA) and can be found in their annual reports, "Global Status of Commercialized Transgenic Crops".

Farmers have widely adopted GM technology. Between 1996 and 2013, the total surface area of land cultivated with GM crops increased by a factor of 100, from 17,000 square kilometers (4,200,000 acres) to 1,750,000 km^2 (432 million acres). 10% of the world's arable land was planted with GM crops in 2010. As of 2011, 11 different transgenic crops were grown commercially on 395 million acres (160 million hectares) in 29 countries such as the US, Brazil, Argentina, India, Canada, China, Paraguay, Pakistan, South Africa, Uruguay, Bolivia, Australia, Philippines, Myanmar, Burkina Faso, Mexico and Spain. One of the key reasons for this widespread adoption is the perceived economic benefit the technology brings to farmers. For example, the system of planting glyphosate-resistant seed and then applying glyphosate once plants emerged provided farmers with the opportunity to dramatically increase the yield from a given plot of land, since this allowed them to plant rows closer together. Without it, farmers had to plant rows far enough apart to control post-emergent weeds with mechanical tillage. Likewise, using Bt seeds means that farmers do not have to purchase insecticides, and then invest time, fuel, and equipment in applying them. However critics have disputed whether yields are higher and whether chemical use is less, with GM crops.

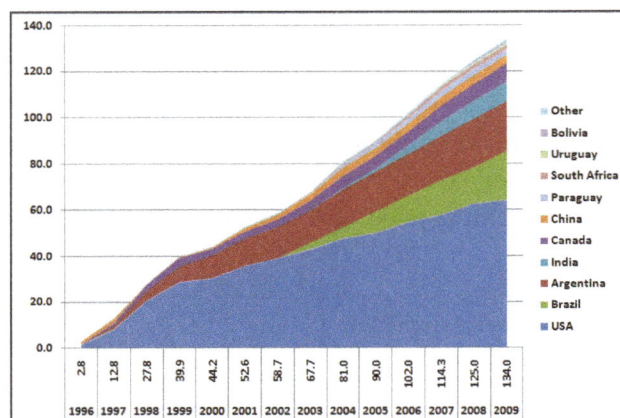

Land area used for genetically modified crops by country, in millions of hectares. In 2011, the land area used was 160 million hectares, or 1.6 million square kilometers.

In the US, by 2014, 94% of the planted area of soybeans, 96% of cotton and 93% of corn were genetically modified varieties. Genetically modified soybeans carried herbicide-tolerant traits only, but maize and cotton carried both herbicide tolerance and insect protection traits (the latter largely Bt protein). These constitute "input-traits" that are aimed to financially benefit the producers, but may have indirect environmental benefits and cost benefits to consumers. The Grocery Manufacturers of America estimated in 2003 that 70–75% of all processed foods in the U.S. contained a GM ingredient.

Europe grows relatively few genetically engineered crops with the exception of Spain, where one fifth of maize is genetically engineered, and smaller amounts in five other countries. The EU had a 'de facto' ban on the approval of new GM crops, from 1999 until 2004. GM crops are now regulated by the EU. In 2015, genetically engineered crops are banned in 38 countries worldwide, 19 of them in Europe. Developing countries grew 54 percent of genetically engineered crops in 2013.

In recent years GM crops expanded rapidly in developing countries. In 2013 approximately 18 million farmers grew 54% of worldwide GM crops in developing countries. 2013's largest increase was in Brazil (403,000 km² versus 368,000 km² in 2012). GM cotton began growing in India in 2002, reaching 110,000 km² in 2013.

"A total of 36 countries (35 + EU-28) have granted regulatory approvals for biotech crops for food and/or feed use and for environmental release or planting since 1994 a total of 2,833 regulatory approvals involving 27 GM crops and 336 GM events (NB: an "event" is a specific genetic modification in a specific species) have been issued by authorities, of which 1,321 are for food use (direct use or processing), 918 for feed use (direct use or processing) and 599 for environmental release or planting. Japan has the largest number (198), followed by the U.S.A. (165, not including "stacked" events), Canada (146), Mexico (131), South Korea (103), Australia (93), New Zealand (83), European Union (71 including approvals that have expired or under renewal process), Philippines (68), Taiwan (65), Colombia (59), China (55) and South Africa (52). Maize has the largest number (130 events in 27 countries), followed by cotton (49 events in 22 countries), potato (31 events in 10 countries), canola (30 events in 12 countries) and soybean (27 events in 26 countries).

Genetically Modified Food

Genetically modified foods (GM foods), also known as genetically engineered foods (GE foods), or bioengineered foods are foods produced from organisms that have had changes introduced into their DNA using the methods of genetic engineering. Genetic engineering techniques allow for the introduction of new traits as well as greater control over traits when compared to previous methods, such as selective breeding and mutation breeding.

Commercial sale of genetically modified foods began in 1994, when Calgene first marketed its unsuccessful Flavr Savr delayed-ripening tomato. Most food modifications have primarily focused on cash crops in high demand by farmers such as soybean, corn, canola, and cotton. Genetically modified crops have been engineered for resistance to pathogens and herbicides and for better nutrient profiles. GM livestock have been developed, although, as of November 2013, none were on the market.

There is a scientific consensus that currently available food derived from GM crops poses no greater risk to human health than conventional food, but that each GM food needs to be tested on a case-by-case basis before introduction. Nonetheless, members of the public are much less likely than scientists to perceive GM foods as safe. The legal and regulatory status of GM foods varies by country, with some nations banning or restricting them, and others permitting them with widely differing degrees of regulation.

However, there are ongoing public concerns related to food safety, regulation, labelling, environmental impact, research methods, and the fact that some GM seeds, along with all new plant varieties, are subject to plant breeders' rights owned by corporations.

Genetically modified foods are foods produced from organisms that have had changes introduced into their DNA using the methods of genetic engineering as opposed to traditional cross breeding. In the U.S., the Department of Agriculture (USDA) and the Food and Drug Administration (FDA) favor the use of the term *genetic engineering* over *genetic modification* as being more precise; the USDA defines *genetic modification* to include "genetic engineering or other more traditional methods".

According to the World Health Organization, "genetically modified organisms (GMOs) can be defined as organisms (i.e. plants, animals or microorganisms) in which the genetic material (DNA) has been altered in a way that does not occur naturally by mating and/or natural recombination. The technology is often called 'modern biotechnology' or 'gene technology', sometimes also 'recombinant DNA technology' or 'genetic engineering'. Foods produced from or using GM organisms are often referred to as GM foods."

Process

Genetically engineered organisms are generated and tested in the laboratory for desired qualities. The most common modification is to add one or more genes to an organism's genome. Less commonly, genes are removed or their expression is increased or silenced or the number of copies of a gene is increased or decreased.

Once satisfactory strains are produced, the producer applies for regulatory approval to field-test them, called a "field release". Field-testing involves cultivating the plants on farm fields or growing animals in a controlled environment. If these field tests are successful, the producer applies for regulatory approval to grow and market the crop. Once approved, specimens (seeds, cuttings, breeding pairs, etc.) are cultivated and sold to farmers. The farmers cultivate and market the new strain. In some cases, the approval covers marketing but not cultivation.

According to the USDA, the number of field releases for genetically engineered organisms has grown from four in 1985 to an average of about 800 per year. Cumulatively, more than 17,000 releases had been approved through September 2013.

Crops

Fruits and Vegetables

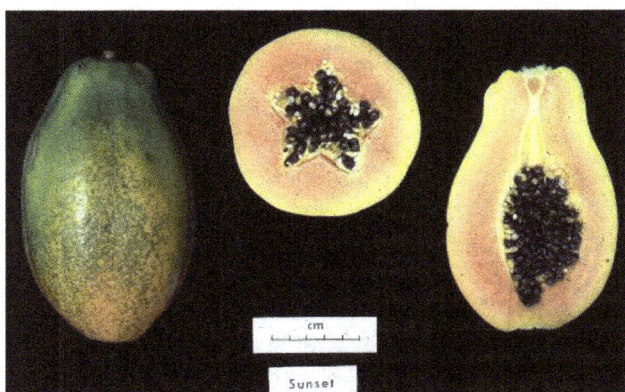

Three views of a papaya, cultivar "Sunset", which was genetically modified to create the cultivar 'SunUp', which is resistant to Papaya ringspot virus.

Papaya was genetically modified to resist the ringspot virus (PSRV). "SunUp" is a transgenic red-fleshed Sunset papaya cultivar that is homozygous for the coat protein gene PRSV; "Rainbow" is a yellow-fleshed F1 hybrid developed by crossing 'SunUp' and nontransgenic yellow-fleshed "Kapoho". "In the early 1990s, Hawaii's papaya industry was facing disaster because of the deadly papaya ringspot virus. Its single-handed savior was a breed engineered to be resistant to the virus. Without it, the state's papaya industry would have collapsed. Today, 80% of Hawaiian papaya is genetically engineered, and there is still no conventional or organic method to control ringspot virus." The GM cultivar was approved in 1998. In China, a transgenic PRSV-resistant papaya was developed by South China Agricultural University and was first approved for commercial planting in 2006; as of 2012 95% of the papaya grown in Guangdong province and 40% of the papaya grown in Hainan province was genetically modified. In Hong Kong, where there is an exemption on growing and releasing any varieties of GM papaya, more than 80% of grown and imported papayas were transgenic.

The New Leaf potato, a GM food developed using naturally occurring bacteria found in the soil known as *Bacillus thuringiensis* (Bt), was made to provide in-plant protection from the yield-robbing Colorado potato beetle. The New Leaf potato, brought to market by Monsanto in the late 1990s, was developed for the fast food market. It was withdrawn in 2001 after retailers rejected it and food processors ran into export problems.

As of 2005, about 13% of the Zucchini (a form of squash) grown in the US was genetically modified to resist three viruses; that strain is also grown in Canada.

Plums genetically engineered for resistance to plum pox, a disease carried by aphids.

In 2011, BASF requested the European Food Safety Authority's approval for cultivation and marketing of its Fortuna potato as feed and food. The potato was made resistant to late blight by adding resistant genes blb1 and blb2 that originate from the Mexican wild potato Solanum bulbocastanum. In February 2013, BASF withdrew its application.

In 2013, the USDA approved the import of a GM pineapple that is pink in color and that "overexpresses" a gene derived from tangerines and suppress other genes, increasing production of lycopene. The plant's flowering cycle was changed to provide for more uniform growth and quality. The fruit "does not have the ability to propagate and persist in the environment once they have been harvested", according to USDA APHIS. According to Del Monte's submission, the pineapples are commercially grown in a "monoculture" that prevents seed production, as the plant's flowers aren't exposed to compatible pollen sources. Importation into Hawaii is banned for "plant sanitation" reasons.

In 2014, the USDA approved a genetically modified potato developed by J. R. Simplot Company that contained ten genetic modifications that prevent bruising and produce less acrylamide when fried. The modifications eliminate specific proteins from the potatoes, via RNA interference, rather than introducing novel proteins.

In February 2015 Arctic Apples were approved by the USDA, becoming the first genetically modified apple approved for sale in the US. Gene silencing is used to reduce the expression of polyphenol oxidase (PPO), thus preventing the fruit from browning.

Corn

Corn used for food and ethanol has been genetically modified to tolerate various herbicides and to express a protein from *Bacillus thuringiensis* (Bt) that kills certain insects. About 90% of the corn grown in the US was genetically modified in 2010. In the US in 2015, 81% of corn acreage contained the Bt trait and 89% of corn acreage contained the glyphosate-tolerant trait. Corn can be processed into grits, meal and flour as an ingredient in pancakes, muffins, doughnuts, breadings and batters, as well as baby foods, meat products, cereals and some fermented products. Corn-based masa flour and masa dough are used in the production of taco shells, corn chips and tortillas.

Soy

Genetically modified soybean has been modified to tolerate herbicides and produce healthier oils. In 2015, 94% of soybean acreage in the U.S. was genetically modified to be glyphosate-tolerant.

Wheat

As of December 2017, genetically modified wheat has been evaluated in field trials, but has not been released commercially.

Derivative Products

Corn Starch and Starch Sugars including Syrups

Starch or amylum is a polysaccharide produced by all green plants as an energy store. Pure starch is a white, tasteless and odourless powder. It consists of two types of molecules: the linear and helical amylose and the branched amylopectin. Depending on the plant, starch generally contains 20 to 25% amylose and 75 to 80% amylopectin by weight.

Starch can be further modified to create modified starch for specific purposes, including creation of many of the sugars in processed foods. They include:

- Maltodextrin, a lightly hydrolyzed starch product used as a bland-tasting filler and thickener.

- Various glucose syrups, also called corn syrups in the US, viscous solutions used as sweeteners and thickeners in many kinds of processed foods.

- Dextrose, commercial glucose, prepared by the complete hydrolysis of starch.

- High fructose syrup, made by treating dextrose solutions with the enzyme glucose isomerase, until a substantial fraction of the glucose has been converted to fructose.

- Sugar alcohols, such as maltitol, erythritol, sorbitol, mannitol and hydrogenated starch hydrolysate, are sweeteners made by reducing sugars.

Lecithin

Lecithin is a naturally occurring lipid. It can be found in egg yolks and oil-producing plants. It is an emulsifier and thus is used in many foods. Corn, soy and safflower oil are sources of lecithin, though the majority of lecithin commercially available is derived from soy. Sufficiently processed lecithin is often undetectable with standard testing practices. According to the FDA, no evidence shows or suggests hazard to the public when lecithin is used at common levels. Lecithin added to foods amounts to only 2 to 10 percent of the 1 to 5 g of phosphoglycerides consumed daily on average. Nonetheless, consumer concerns about GM food extend to such products. This concern led to policy and regulatory changes in Europe in 2000, when Regulation (EC) 50/2000 was passed which required labelling of food containing additives derived from GMOs, including lecithin. Because of the difficulty of detecting the origin of derivatives like lecithin with current testing practices, European regulations require those who wish to sell lecithin in Europe to employ a comprehensive system of Identity preservation (IP).

Sugar

The US imports 10% of its sugar, while the remaining 90% is extracted from sugar beet and sugarcane. After deregulation in 2005, glyphosate-resistant sugar beet was extensively adopted in the United States. 95% of beet acres in the US were planted with glyphosate-resistant seed in 2011. GM sugar beets are approved for cultivation in the US, Canada and Japan; the vast majority are grown in the US. GM beets are approved for import and consumption in Australia, Canada, Colombia, EU, Japan, Korea, Mexico, New Zealand, Philippines, the Russian Federation and Singapore. Pulp from the refining process is used as animal feed. The sugar produced from GM sugar beets contains no DNA or protein – it is just sucrose that is chemically indistinguishable from sugar produced from non-GM sugar beets. Independent analyses conducted by internationally recognized laboratories found that sugar from Roundup Ready sugar beets is identical to the sugar from comparably grown conventional (non-Roundup Ready) sugar beets.

Vegetable Oil

Most vegetable oil used in the US is produced from GM crops canola, corn, cotton and soybeans. Vegetable oil is sold directly to consumers as cooking oil, shortening and margarine and is used in prepared foods. There is a vanishingly small amount of protein or DNA from the original crop in vegetable oil. Vegetable oil is made of triglycerides extracted from plants or seeds and then refined and may be further processed via hydrogenation to turn liquid oils into solids. The refining process removes all, or nearly all non-triglyceride ingredients. Medium-chain triglycerides (MCTs) offer an alternative to conventional fats and oils. The length of a fatty acid influences its fat absorption during the digestive process. Fatty acids in the middle position on the glycerol molecules appear to be absorbed more easily and influence metabolism more than fatty acids on the end positions. Unlike ordinary fats, MCTs are metabolized like carbohydrates. They have exceptional oxidative stability, and prevent foods from turning rancid readily.

Other Uses

Animal Feed

Livestock and poultry are raised on animal feed, much of which is composed of the leftovers from processing crops, including GM crops. For example, approximately 43% of a canola seed is oil. What remains after oil extraction is a meal that becomes an ingredient in animal feed and contains canola protein. Likewise, the bulk of the soybean crop is grown for oil and meal. The high-protein defatted and toasted soy meal becomes livestock feed and dog food. 98% of the US soybean crop goes for livestock feed. In 2011, 49% of the US maize harvest was used for livestock feed (including the percentage of waste from distillers grains). "Despite methods that are becoming more and more sensitive, tests have not yet been able to establish a difference in the meat, milk, or eggs of animals depending on the type of feed they are fed. It is impossible to tell if an animal was fed GM soy just by looking at the resulting meat, dairy, or egg products. The only way to verify the presence of GMOs in animal feed is to analyze the origin of the feed itself."

A 2012 literature review of studies evaluating the effect of GM feed on the health of animals did not find evidence that animals were adversely affected, although small biological differences were occasionally found. The studies included in the review ranged from 90 days to two years, with several of the longer studies considering reproductive and intergenerational effects.

Enzymes produced by genetically modified microorganisms are also integrated into animal feed to enhance availability of nutrients and overall digestion. These enzymes may also provide benefit to the gut microbiome of an animal, as well as hydrolyse antinutritional factors present in the feed.

Proteins

Rennet is a mixture of enzymes used to coagulate milk into cheese. Originally it was available only from the fourth stomach of calves, and was scarce and expensive, or was available from microbial sources, which often produced unpleasant tastes. Genetic engineering made it possible to extract rennet-producing genes from animal stomachs and insert them into bacteria, fungi or yeasts to make them produce chymosin, the key enzyme. The modified microorganism is killed after fermentation. Chymosin is isolated from the fermentation broth, so that the Fermentation-Produced Chymosin (FPC) used by cheese producers has an amino acid sequence that is identical to bovine rennet. The majority of the applied chymosin is retained in the whey. Trace quantities of chymosin may remain in cheese.

FPC was the first artificially produced enzyme to be approved by the US Food and Drug Administration. FPC products have been on the market since 1990 and as of 2015 had yet to be surpassed in commercial markets. In 1999, about 60% of US hard cheese was made with FPC. Its global market share approached 80%. By 2008, approximately 80% to 90% of commercially made cheeses in the US and Britain were made using FPC.

In some countries, recombinant (GM) bovine somatotropin (also called rBST, or bovine growth hormone or BGH) is approved for administration to increase milk production. rBST may be present in milk from rBST treated cows, but it is destroyed in the digestive system and even if directly injected into the human bloodstream, has no observable effect on humans. The FDA, World Health Organization, American Medical Association, American Dietetic Association and the National Institutes of Health have independently stated that dairy products and meat from rBST-treated cows are safe for human consumption. However, on 30 September 2010, the United States Court of Appeals, Sixth Circuit, analyzing submitted evidence, found a "compositional difference" between milk from rBGH-treated cows and milk from untreated cows. The court stated that milk from rBGH-treated cows has: increased levels of the hormone Insulin-like growth factor 1 (IGF-1); higher fat content and lower protein content when produced at certain points in the cow's lactation cycle; and more somatic cell counts, which may "make the milk turn sour more quickly".

Livestock

Genetically modified livestock are organisms from the group of cattle, sheep, pigs, goats, birds, horses and fish kept for human consumption, whose genetic material (DNA) has been altered using genetic engineering techniques. In some cases, the aim is to introduce a new trait to the animals which does not occur naturally in the species, i.e. transgenesis.

A 2003 review published on behalf of Food Standards Australia New Zealand examined transgenic experimentation on terrestrial livestock species as well as aquatic species such as fish and shellfish. The review examined the molecular techniques used for experimentation as well as techniques for tracing the transgenes in animals and products as well as issues regarding transgene stability.

Some mammals typically used for food production have been modified to produce non-food products, a practice sometimes called Pharming.

Salmon

A GM salmon, awaiting regulatory approval since 1997, was approved for human consumption by the American FDA in November 2015, to be raised in specific land-based hatcheries in Canada and Panama.

Health and Safety

There is a scientific consensus that currently available food derived from GM crops poses no greater risk to human health than conventional food, but that each GM food needs to be tested on a case-by-case basis before introduction. Nonetheless, members of the public are much less likely than scientists to perceive GM foods as safe. The legal and regulatory status of GM foods varies by country, with some nations banning or restricting them, and others permitting them with widely differing degrees of regulation.

Opponents claim that long-term health risks have not been adequately assessed and propose various combinations of additional testing, labeling or removal from the market. The advocacy group European Network of Scientists for Social and Environmental Responsibility (ENSSER), disputes the claim that "science" supports the safety of current GM foods, proposing that each GM food must be judged on case-by-case basis.

Testing

The legal and regulatory status of GM foods varies by country, with some nations banning or restricting them, and others permitting them with widely differing degrees of regulation. Countries such as the United States, Canada, Lebanon and Egypt use *substantial equivalence* to determine if further testing is required, while many countries such as those in the European Union, Brazil and China only authorize GMO cultivation on a case-by-case basis. In the U.S. the FDA determined that GMO's are "Generally Recognized as Safe" (GRAS) and therefore do not require additional testing if the GMO product is substantially equivalent to the non-modified product. If new substances are found, further testing may be required to satisfy concerns over potential toxicity, allergenicity, possible gene transfer to humans or genetic outcrossing to other organisms.

Regulation

Government regulation of GMO development and release varies widely between countries. Marked differences separate GMO regulation in the U.S. and GMO regulation in the European Union.

Regulation also varies depending on the intended product's use. For example, a crop not intended for food use is generally not reviewed by authorities responsible for food safety.

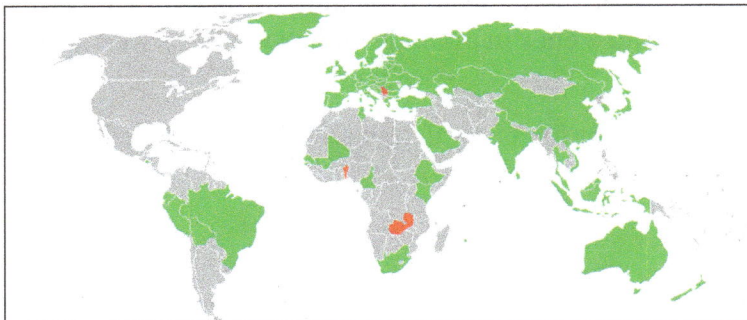

Green: Mandatory labeling required; Red: Ban on import and cultivation of genetically engineered food.

United States Regulations

In the U.S., three government organizations regulate GMOs. The FDA checks the chemical composition of organisms for potential allergens. The United States Department of Agriculture (USDA) supervises field testing and monitors the distribution of GM seeds. The United States Environmental Protection Agency (EPA) is responsible for monitoring pesticide usage, including plants modified to contain proteins toxic to insects. Like USDA, EPA also oversees field testing and the distribution of crops that have had contact with pesticides to ensure environmental safety. In 2015 the Obama administration announced that it would update the way the government regulated GM crops.

In 1992 FDA published "Statement of Policy: Foods derived from New Plant Varieties". This statement is a clarification of FDA's interpretation of the Food, Drug, and Cosmetic Act with respect to foods produced from new plant varieties developed using recombinant deoxyribonucleic acid (rDNA) technology. FDA encouraged developers to consult with the FDA regarding any bioengineered foods in development. The FDA says developers routinely do reach out for consultations. In 1996 FDA updated consultation procedures.

The StarLink corn recalls occurred in the autumn of 2000, when over 300 food products were found to contain a genetically modified corn that had not been approved for human consumption. It was the first-ever recall of a genetically modified food.

Labeling

As of 2015, 64 countries require labeling of GMO products in the marketplace.

US and Canadian national policy is to require a label only given significant composition differences or documented health impacts, although some individual US states (Vermont, Connecticut and Maine) enacted laws requiring them. In July 2016, Public Law 114-214 was enacted to regulate labeling of GMO food on a national basis.

In some jurisdictions, the labeling requirement depends on the relative quantity of GMO in the product. A study that investigated voluntary labeling in South Africa found that 31% of products labeled as GMO-free had a GM content above 1.0%.

In the European Union all food (including processed food) or feed that contains greater than 0.9% GMOs must be labelled.

Detection

Testing on GMOs in food and feed is routinely done using molecular techniques such as PCR and bioinformatics.

In a January 2010 paper, the extraction and detection of DNA along a complete industrial soybean oil processing chain was described to monitor the presence of Roundup Ready (RR) soybean: "The amplification of soybean lectin gene by end-point polymerase chain reaction (PCR) was successfully achieved in all the steps of extraction and refining processes, until the fully refined soybean oil. The amplification of RR soybean by PCR assays using event-specific primers was also achieved for all the extraction and refining steps, except for the intermediate steps of refining (neutralisation, washing and bleaching) possibly due to sample instability. The real-time PCR assays using specific probes confirmed all the results and proved that it is possible to detect and quantify genetically modified organisms in the fully refined soybean oil. To our knowledge, this has never been reported before and represents an important accomplishment regarding the traceability of genetically modified organisms in refined oils."

According to Thomas Redick, detection and prevention of cross-pollination is possible through the suggestions offered by the Farm Service Agency (FSA) and Natural Resources Conservation Service (NRCS). Suggestions include educating farmers on the importance of coexistence, providing farmers with tools and incentives to promote coexistence, conduct research to understand and monitor gene flow, provide assurance of quality and diversity in crops, provide compensation for actual economic losses for farmers.

Controversies

The genetically modified foods controversy consists of a set of disputes over the use of food made from genetically modified crops. The disputes involve consumers, farmers, biotechnology companies, governmental regulators, non-governmental organizations, environmental and political activists and scientists. The major disagreements include whether GM foods can be safely consumed, harm the environment and/or are adequately tested and regulated. The objectivity of scientific research and publications has been challenged. Farming-related disputes include the use and impact of pesticides, seed production and use, side effects on non-GMO crops/farms, and potential control of the GM food supply by seed companies.

The conflicts have continued since GM foods were invented. They have occupied the media, the courts, local, regional, national governments, and international organizations.

Genetically Modified Organism

Genetically modified organism (GMO) is an organism whose genome has been engineered in the laboratory in order to favour the expression of desired physiological traits or the production of desired biological products. In conventional livestock production, crop farming, and even pet breeding, it has long been the practice to breed select individuals of a species in order to produce offspring that have desirable traits. In genetic modification, however, recombinant genetic technologies are employed to produce organisms whose genomes have been precisely altered at the molecular level, usually by the inclusion of genes from unrelated species of organisms that code for traits that would not be obtained easily through conventional selective breeding.

GMOs are produced through using scientific methods that include recombinant DNA technology and reproductive cloning. In reproductive cloning, a nucleus is extracted from a cell of the individual to be cloned and is inserted into the enucleated cytoplasm of a host egg. The process results in the generation of an offspring that is genetically identical to the donor individual. The first animal produced by means of this cloning technique with a nucleus from an adult donor cell (as opposed to a donor embryo) was a sheep named Dolly, born in 1996. Since then a number of other animals, including pigs, horses, and dogs, have been generated by reproductive cloning technology. Recombinant DNA technology, on the other hand, involves the insertion of one or more individual genes from an organism of one species into the DNA (deoxyribonucleic acid) of another. Whole-genome replacement, involving the transplantation of one bacterial genome into the "cell body," or cytoplasm, of another microorganism, has been reported, although this technology is still limited to basic scientific applications.

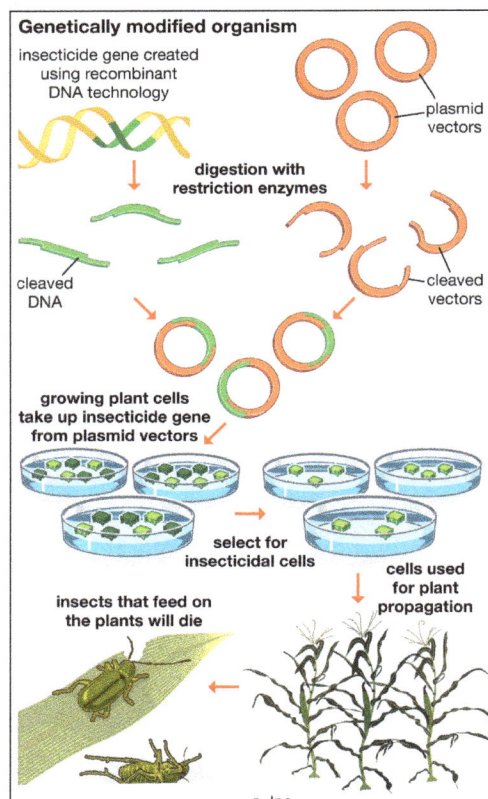

Genetically modified organisms are produced using scientific methods that include recombinant DNA technology.

GMOs produced through genetic technologies have become a part of everyday life, entering into society through agriculture, medicine, research, and environmental management. However, while GMOs have benefited human society in many ways, some disadvantages exist; therefore, the production of GMOs remains a highly controversial topic in many parts of the world.

GMOs in Agriculture

Genetically modified (GM) foods were first approved for human consumption in the United States in 1994, and by 2014–15 about 90 percent of the corn, cotton, and soybeans planted in the United States were GM. By the end of 2010, GM crops covered more than 10 million square kilometres (3.86 million square miles) of land in 29 countries worldwide—one-tenth of the world's farmland. The majority of GM crops were grown in the Americas.

Genetically modified corn (maize).

Engineered crops can dramatically increase per area crop yields and, in some cases, reduce the use of chemical insecticides. For example, the application of wide-spectrum insecticides declined in many areas growing plants, such as potatoes, cotton, and corn, that were endowed with a gene from the bacterium Bacillus thuringiensis, which produces a natural insecticide called Bt toxin. Field studies conducted in India in which Bt cotton was compared with non-Bt cotton demonstrated a 30–80 percent increase in yield from the GM crop. This increase was attributed to marked improvement in the GM plants' ability to overcome bollworm infestation, which was otherwise common. Studies of Bt cotton production in Arizona, U.S., demonstrated only small gains in yield—about 5 percent—with an estimated cost reduction of $25–$65 (USD) per acre owing to decreased pesticide applications. In China, where farmers first gained access to Bt cotton in 1997, the GM crop was initially successful. Farmers who had planted Bt cotton reduced their pesticide use by 50–80 percent and increased their earnings by as much as 36 percent. By 2004, however, farmers who had been growing Bt cotton for several years found that the benefits of the crop eroded as populations of secondary insect pests, such as mirids, increased. Farmers once again were forced to spray broad-spectrum pesticides throughout the growing season, such that the average revenue for Bt growers was 8 percent lower than that of farmers who grew conventional cotton. Meanwhile, Bt resistance had also evolved in field populations of major cotton pests, including both the cotton bollworm (Helicoverpa armigera) and the pink bollworm (Pectinophora gossypiella).

Other GM plants were engineered for resistance to a specific chemical herbicide, rather than resistance to a natural predator or pest. Herbicide-resistant crops (HRC) have been available since the mid-1980s; these crops enable effective chemical control of weeds, since only the HRC plants can survive in fields treated with the corresponding herbicide. Many HRCs are resistant to glyphosate (Roundup), enabling liberal application of the chemical, which is highly effective against weeds. Such crops have been especially valuable for no-till farming, which helps prevent soil erosion. However, because HRCs encourage increased application of chemicals to the soil, rather than decreased application, they remain controversial with regard to their environmental impact. In addition, in order to reduce the risk of selecting for herbicide-resistant weeds, farmers must use multiple diverse weed-management strategies.

Another example of a GM crop is "golden" rice, which originally was intended for Asia and was genetically modified to produce almost 20 times the beta-carotene of previous varieties. Golden rice was created by modifying the rice genome to include a gene from the daffodil Narcissus pseudonarcissus that produces an enzyme known as phyotene synthase and a gene from the bacterium Erwinia uredovora that produces an enzyme called phyotene desaturase. The introduction of these genes enabled beta-carotene, which is converted to vitamin A in the human liver, to accumulate in the rice endosperm—the edible part of the rice plant—thereby increasing the amount of beta-carotene available for vitamin A synthesis in the body. In 2004 the same researchers who developed the original golden rice plant improved upon the model, generating golden rice 2, which showed a 23-fold increase in carotenoid production.

Another form of modified rice was generated to help combat iron deficiency, which impacts close to 30 percent of the world population. This GM crop was engineered by introducing into the rice genome a ferritin gene from the common bean, Phaseolus vulgaris, that produces a protein capable of binding iron, as well as a gene from the fungus Aspergillus fumigatus that produces an enzyme capable of digesting compounds that increase iron bioavailability via digestion of phytate (an inhibitor of iron absorption). The iron-fortified GM rice was engineered to overexpress an existing rice gene that produces a cysteine-rich metallothioneinlike (metal-binding) protein that enhances iron absorption.

A variety of other crops modified to endure the weather extremes common in other parts of the globe are also in production.

GMOs in Medicine and Research

GMOs have emerged as one of the mainstays of biomedical research since the 1980s. For example, GM animal models of human genetic diseases enabled researchers to test novel therapies and to explore the roles of candidate risk factors and modifiers of disease outcome. GM microbes, plants, and animals also revolutionized the production of complex pharmaceuticals by enabling the generation of safer and cheaper vaccines and therapeutics. Pharmaceutical products range from recombinant hepatitis B vaccine produced by GM baker's yeast to injectable insulin (for diabetics) produced in GM Escherichia coli bacteria and to factor VIII (for hemophiliacs) and tissue plasminogen activator (tPA, for heart attack or stroke patients), both of which are produced in GM mammalian cells grown in laboratory culture. Furthermore, GM plants that produce "edible vaccines" are under development. An edible vaccine is an antigenic protein that is produced in the consumable parts of a plant (e.g., fruit) and absorbed into the bloodstream when the parts are eaten. Once absorbed into the body, the protein stimulates the immune system to produce antibodies against the pathogen from which the antigen was derived. Such vaccines could offer a safe, inexpensive, and painless way to provide vaccines, particularly in less-developed regions of the world, where the limited availability of refrigeration and sterile needles has been problematic for some traditional vaccines. Novel DNA vaccines may be useful in the struggle to prevent diseases that have proved resistant to traditional vaccination approaches, including HIV/AIDS, tuberculosis, and cancer.

Genetic modification of insects has become an important area of research, especially in the struggle to prevent parasitic diseases. For example, GM mosquitoes have been developed that express a small protein called SM1, which blocks entry of the malaria parasite, Plasmodium, into the mosquito's gut. This results in the disruption of the parasite's life cycle and renders the mosquito malaria-resistant. Introduction of these GM mosquitoes into the wild could help reduce transmission of the malaria parasite. In another example, male Aedes aegypti mosquitoes engineered with a method known as the sterile insect technique transmit a gene to their offspring that causes the offspring to die before becoming sexually mature. In field trials in a Brazil suburb, A. aegypti populations declined by 95 percent following the sustained release of sterile GM males.

Finally, genetic modification of humans via gene therapy is becoming a treatment option for diseases ranging from rare metabolic disorders to cancer. Coupling stem cell technology with recombinant DNA methods allows stem cells derived from a patient to be modified in the laboratory to introduce a desired gene. For example, a normal beta-globin gene may be introduced into the DNA of bone marrow-derived hematopoietic stem cells from a patient with sickle cell anemia; introduction of these GM cells into the patient could cure the disease without the need for a matched donor.

Role of GMOs in Environmental Management

Another application of GMOs is in the management of environmental issues. For example, some bacteria can produce biodegradable plastics, and the transfer of that ability to microbes that can be easily grown in the laboratory may enable the wide-scale "greening" of the plastics industry. In the early 1990s, Zeneca, a British company, developed a microbially produced biodegradable plastic called Biopol (polyhydroxyalkanoate, or PHA). The plastic was made with the use of a GM bacterium, Ralstonia eutropha, to convert glucose and a variety of organic acids into a flexible polymer. GMOs endowed with the bacterially encoded ability to metabolize oil and heavy metals may provide efficient bioremediation strategies.

Sociopolitical Relevance of GMOs

While GMOs offer many potential benefits to society, the potential risks associated with them have fueled controversy, especially in the food industry. Many skeptics warn about the dangers that GM crops may pose to human health. For example, genetic manipulation may potentially alter the allergenic properties of crops. Whether some GM crops, such as golden rice, deliver on the promise of improved health benefits is also unclear. The release of GM mosquitoes and other GMOs into the environment also raised concerns. More-established risks were associated with the potential spread of engineered crop genes to native flora and the possible evolution of insecticide-resistant "superbugs."

From the late 1990s, the European Union (EU) addressed such concerns by implementing strict GMO labeling laws. In the early 2000s, all GM foods and GM animal feeds in the EU were required to be labeled if they consisted of or contained GM products in a proportion greater than 0.9 percent. By contrast, in the United States, foods containing GM ingredients did not require special labeling, though the issue was hotly debated at national and state levels. Many opponents of GM products focused their arguments on unknown risks to food safety. However, despite the concerns of some consumer and health groups, especially in Europe, numerous scientific panels, including the U.S. Food and Drug Administration, concluded that consumption of GM foods was safe, even in cases involving GM foods with genetic material from very distantly related organisms.

The strict regulations on GM products in the EU have been a source of tension in agricultural trade.

In the late 1990s, the EU declared a moratorium on the import and use of GM crops. However, the ban—which led to trade disputes with other countries, particularly the United States, where GM foods had been accepted openly—was considered unjustified by the World Trade Organization. In consequence, the EU implemented regulatory changes that allowed for the import of certain GM crops. Within Europe, however, only one GM crop, a type of insect-resistant corn (maize), was cultivated. Some countries, including certain African states, had likewise rejected GM products. Still other countries, such as Canada, China, Argentina, and Australia, had open policies on GM foods.

The use of GMOs in medicine and research has produced a debate that is more philosophical in nature. For example, while genetic researchers believe they are working to cure disease and ameliorate suffering, many people worry that current gene therapy approaches may one day be applied to produce "designer" children or to lengthen the natural human life span. Similar to many other technologies, gene therapy and the production and application of GMOs can be used to address and resolve complicated scientific, medical, and environmental issues, but they must be used wisely.

Biomimetics

The tiny hooks on bur fruits (left) inspired Velcro tape (right).

Biomimetics or biomimicry is the imitation of the models, systems, and elements of nature for the purpose of solving complex human problems. A closely related field is bionics.

Living organisms have evolved well-adapted structures and materials over geological time through natural selection. Biomimetics has given rise to new technologies inspired by biological solutions at macro and nanoscales. Humans have looked at nature for answers to problems throughout our existence. Nature has solved engineering problems such as self-healing abilities, environmental exposure tolerance and resistance, hydrophobicity, self-assembly, and harnessing solar energy.

Bio-inspired Technologies

Biomimetics could in principle be applied in many fields. Because of the diversity and complexity of biological systems, the number of features that might be imitated is large. Biomimetic applications are at various stages of development from technologies that might become commercially usable to prototypes. Murray's law, which in conventional form determined the optimum diameter of blood vessels, has been re-derived to provide simple equations for the pipe or tube diameter which gives a minimum mass engineering system.

Locomotion

Aircraft wing design and flight techniques are being inspired by birds and bats. Biorobots based on the physiology and methods of locomotion of animals include BionicKangaroo which moves like a kangaroo, saving energy from one jump and transferring it to its next jump. Kamigami Robots, a children's toy, mimic cockroach locomotion to run quickly and efficiently over indoor and outdoor surfaces.

Construction and Architecture

Researchers studied the termite's ability to maintain virtually constant temperature and humidity in their termite mounds in Africa despite outside temperatures that vary from 1.5 °C to 40 °C (35 °F to 104 °F). Researchers initially scanned a termite mound and created 3-D images of the mound structure, which revealed construction that could influence human building design. The Eastgate Centre, a mid-rise office complex in Harare, Zimbabwe, stays cool without air conditioning and uses only 10% of the energy of a conventional building of the same size.

In structural engineering, the Swiss Federal Institute of Technology (EPFL) has incorporated bio-mimetic characteristics in an adaptive deployable "tensegrity" bridge. The bridge can carry out self-diagnosis and self-repair. The arrangement of leaves on a plant has been adapted for better solar power collection.

Analysis of the elastic deformation happening when a pollinisator lands on the sheath-like perch part of the flower *Strelitzia reginae* (known as Bird-of-Paradise flower) has inspired architects and scientists from the University of Freiburg and University of Stuttgart for the creation of hingeless shading systems that can react to their environment. These bio-inspired products is sold under the name Flectofin.

Other hingeless bioinspired system includes Flectofold. Flectofold has been inspired from the trapping system developed by the carnivorous plant *Aldrovanda vesiculosa*.

Structural Materials

There is a great need for new structural materials that are light weight but offer exceptional combinations of stiffness, strength, and toughness.

Such materials would need to be manufactured into bulk materials with complex shapes at high volume and low cost and would serve a variety of fields such as construction, transportation, energy storage and conversion. In a classic design problem, strength and toughness are more likely to be mutually exclusive i.e., strong materials are brittle and tough materials are weak. However, natural materials with complex and hierarchical material gradients that span from nano- to macro-scales are both strong and tough. Generally, most natural materials utilize limited chemical components but complex material architectures that give rise to exceptional mechanical properties. Understanding the highly diverse and multi functional biological materials and discovering approaches to replicate such structures will lead to advanced and more efficient technologies. Bone, nacre (abalone shell), teeth, the dactyl clubs of stomatopod shrimps and bamboo are great examples of damage tolerant materials. The exceptional resistance to fracture of bone is due to complex deformation and toughening mechanisms that operate at spanning different size scales - nanoscale structure of protein molecules to macroscopic physiological scale.

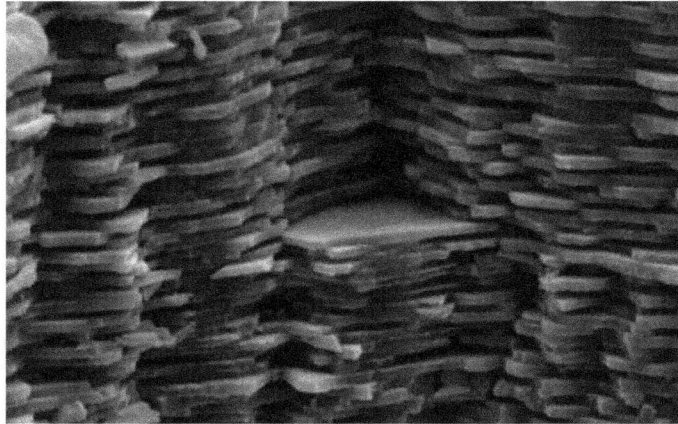

Electron microscopy image of a fractured surface of nacre.

Nacre exhibits similar mechanical properties however with rather simpler structure. Nacre shows a brick and mortar llike structure with thick mineral layer (0.2~0.9-μm) of closely packed aragonite structures and thin organic matrix (~20-nm). While thin films and micrometer sized samples that mimic these structures are already produced, successful production of bulk biomimetic structural materials is yet to be realized. However, numerous processing techniques have been proposed for producing nacre like materials.

Biomorphic mineralization is a technique that produces materials with morphologies and structures resembling those of natural living organisms by using bio-structures as templates for mineralization. Compared to other methods of material production, biomorphic mineralization is facile, environmentally benign and economic.

Freeze casting (Ice templating), an inexpensive method to mimic natural layered structures was employed by researchers at Lawrence Berkeley National Laboratory to create alumina-Al-Si and IT HAP-epoxy layered composites that match the mechanical properties of bone with an equivalent mineral/ organic content. Various further studies also employed similar methods to produce high strength and high toughness composites involving a variety of constituent phases.

Recent studies demonstrated production of cohesive and self supporting macroscopic tissue constructs that mimic living tissues by printing tens of thousands of heterologous picoliter droplets in software-defined, 3D millimeter-scale geometries. Efforts are also taken up to mimic the design of nacre in artificial composite materials using fused deposition modelling and the helicoidal structures of stomatopod clubs in the fabrication of high performance carbon fiber-epoxy composites.

Various established and novel additive manufacturing technologies like PolyJet printing, direct ink writing, 3D magnetic printing, multi-material magnetically assisted 3D printing and magnetically-assisted slip casting have also been utilized to mimic the complex micro-scale architectures of natural materials and provide huge scope for future research.

Spider web silk is as strong as the Kevlar used in bulletproof vests. Engineers could in principle use such a material, if it could be reengineered to have a long enough life, for parachute lines, suspension bridge cables, artificial ligaments for medicine, and other purposes. The self-sharpening teeth of many animals have been copied to make better cutting tools.

New ceramics that exhibit giant electret hysteresis have also been realized.

Self Healing Materials

In general in biological systems, self healing occurs via chemical signals released at the site of fracture which initiate a systemic response that transport repairing agents to the fracture site thereby promoting autonomic healing. To demonstrate the use of micro-vascular networks for autonomic healing, researchers developed a microvascular coating–substrate architecture that mimics human skin. Bio-inspired self-healing structural color hydrogels that maintain the stability of an inverse opal structure and its resultant structural colors were developed. A self-repairing membrane inspired by rapid self-sealing processes in plants was developed for inflatable light weight structures such as rubber boats or Tensairity constructions. The researchers applied a thin soft cellular polyurethane foam coating on the inside of a fabric substrate, which closes the crack if the membrane is punctured with a spike. Self-healing materials, polymers and composite materials capable of mending cracks have been produced based on biological materials.

Surfaces

Surfaces that recreate properties of shark skin are intended to enable more efficient movement through water. Efforts have been made to produce fabric that emulates shark skin.

Surface tension biomimetics are being researched for technologies such as hydrophobic or hydrophilic coatings and microactuators.

Adhesion

Wet Adhesion

Some amphibians, such as tree and torrent frogs and arboreal salamanders, are able to attach to and move over wet or even flooded environments without falling. This kind of organisms have toe pads which are permanently wetted by mucus secreted from glands that open into the channels between epidermal cells. They attach to mating surfaces by wet adhesion and they are capable of climbing on wet rocks even when water is flowing over the surface. Tire treads have also been inspired by the toe pads of tree frogs.

Marine mussels can stick easily and efficiently to surfaces underwater under the harsh conditions of the ocean. Mussels use strong filaments to adhere to rocks in the inter-tidal zones of wave-swept beaches, preventing them from being swept away in strong sea currents. Mussel foot proteins attach the filaments to rocks, boats and practically any surface in nature including other mussels. These proteins contain a mix of amino acid residues which has been adapted specifically for adhesive purposes. Researchers from the University of California Santa Barbara borrowed and simplified chemistries that the mussel foot uses to overcome this engineering challenge of wet adhesion to create copolyampholytes, and one-component adhesive systems with potential for employment in nanofabrication protocols. Other research has proposed adhesive glue from mussels.

Dry Adhesion

Leg attachment pads of several animals, including many insects (e.g. beetles and flies), spiders and lizards (e.g. geckos), are capable of attaching to a variety of surfaces and are used for locomotion, even on vertical walls or across ceilings. Attachment systems in these organisms have similar structures at their terminal elements of contact, known as setae. Such biological examples have offered inspiration in order to produce climbing robots, boots and tape. Synthetic setae have also been developed for the production of dry adhesives.

Optics

Biomimetic materials are gaining increasing attention in the field of optics and photonics. There are still little known bioinspired or biomimetic products involving the photonic properties of plants or animals. However, understanding how nature designed such optical materials from biological resources is worth pursuing and might lead to future commercial products.

Inspiration from Fruits and Plants

For instance, the chiral self-assembly of cellulose inspired by the *Pollia condensata* berry has been exploited to make optically active films. Such films are made from cellulose which is a biodegradable and biobased resource obtained from wood or cotton. The structural colours can potentially be everlasting and have more vibrant colour than the ones obtained from chemical absorption of light. *Pollia condensata* is not the only fruit showing a structural coloured skin, other berries such as *Margaritaria nobilis* does. These fruits show iridescent colors in the blue-green region of the visible spectrum which gives the fruit a strong metallic and shiny visual appearance. The structural colours come from the organisation of cellulose chains in the fruit's epicarp, a part of the fruit skin. Each cell of the epicarp is made of a multilayered envelope that behaves like a Bragg reflector. However, the light which is reflected from the skin of these fruits is not polarised unlike the one arising from man-made replicates obtained from the self-assembly of cellulose nanocrystals into helicoids, which only reflect left-handed circularly polarised light.

The fruit of *Elaeocarpus angustifolius* also show structural colour that come arises from the presence of specialised cells called iridosomes which have layered structures. Similar iridosomes have also been found in Delarbrea michieana fruits.

In plants, multi layer structures can be found either at the surface of the leaves (on top of the epidermis), such as in *Selaginella willdenowii* or within specialized intra-cellular organelles, the so-called iridoplasts, which are located inside the cells of the upper epidermis. For instance, the rain forest plants Begonia pavonina have iridoplasts located inside the epidermal cells.

Structural colours have also been found in several algae, such as in the red alga *Chondrus crispus* (Irish Moss).

Inspiration from Animals

Vibrant blue color of Morpho butterfly due to structural coloration has been mimicked by a variety of technologies.

Structural coloration produces the rainbow colours of soap bubbles, butterfly wings and many beetle scales. Phase-separation has been used to fabricate ultra-white scattering membranes from polymethylmethacrylate, mimicking the beetle *Cyphochilus*. LED lights can be designed to mimic the patterns of scales on fireflies' abdomens, improving their efficiency.

Morpho butterfly wings are structurally coloured to produce a vibrant blue that does not vary with angle. This effect can be mimicked by a variety of technologies. Lotus Cars claim to have developed a paint that mimics the *Morpho* butterfly's structural blue colour. In 2007, Qualcomm commercialised an interferometric modulator display technology, "Mirasol", using *Morpho*-like optical interference. In 2010, the dressmaker Donna Sgro made a dress from Teijin Fibers' Morphotex, an undyed fabric woven from structurally coloured fibres, mimicking the microstructure of *Morpho* butterfly wing scales.

Agricultural Systems

Holistic planned grazing, using fencing and/or herders, seeks to restore grasslands by carefully planning movements of large herds of livestock to mimic the vast herds found in nature. The natural system being mimicked and used as a template is grazing animals concentrated by pack predators that must move on after eating, trampling, and manuring an area, and returning only after it has fully recovered. Developed by Allan Savory, who in turn was inspired by the work of André Voisin, this method of grazing holds tremendous potential in building soil, increasing biodiversity, reversing desertification, and mitigating global warming, similar to what occurred during the past 40 million years as the expansion of grass-grazer ecosystems built deep grassland soils, sequestering carbon and cooling the planet.

Permaculture is a set of design principles centered around whole systems thinking, simulating or directly utilizing the patterns and resilient features observed in natural ecosystems. It uses these principles in a growing number of fields from regenerative agriculture, rewilding, community, and organizational design and development.

Other Uses

Some air conditioning systems use biomimicry in their fans to increase airflow while reducing power consumption.

Other Technologies

Protein folding has been used to control material formation for self-assembled functional nano-structures. Polar bear fur has inspired the design of thermal collectors and clothing. The light refractive properties of the moth's eye has been studied to reduce the reflectivity of solar panels.

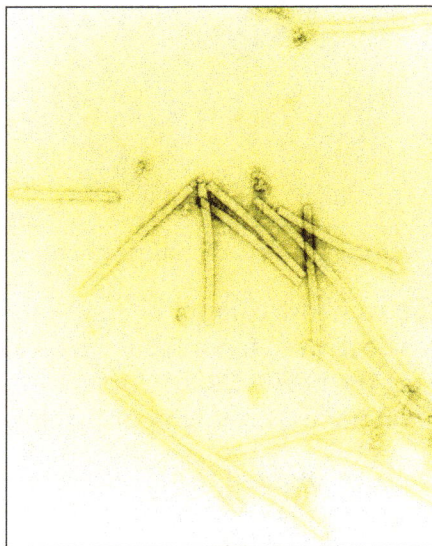

Scanning electron micrograph of rod shaped tobacco mosaic virus particles.

The Bombardier beetle's powerful repellent spray inspired a Swedish company to develop a "micro mist" spray technology, which is claimed to have a low carbon impact (compared to aerosol sprays). The beetle mixes chemicals and releases its spray via a steerable nozzle at the end of its abdomen, stinging and confusing the victim.

Most viruses have an outer capsule 20 to 300 nm in diameter. Virus capsules are remarkably robust and capable of withstanding temperatures as high as 60 °C; they are stable across the pH range 2-10. Viral capsules can be used to create nano device components such as nanowires, nanotubes, and quantum dots. Tubular virus particles such as the tobacco mosaic virus (TMV) can be used as templates to create nanofibers and nanotubes, since both the inner and outer layers of the virus are charged surfaces which can induce nucleation of crystal growth. This was demonstrated through the production of platinum and gold nanotubes using TMV as a template. Mineralized virus particles have been shown to withstand various pH values by mineralizing the viruses with different materials such as silicon, PbS, and CdS and could therefore serve as a useful carriers of material. A spherical plant virus called cowpea chlorotic mottle virus (CCMV) has interesting expanding properties when exposed to environments of pH higher than 6.5. Above this pH, 60 independent pores with diameters about 2 nm begin to exchange substance with the environment. The structural transition of the viral capsid can be utilized in Biomorphic mineralization for selective uptake and deposition of minerals by controlling the solution pH. Possible applications include using the viral cage to produce uniformly shaped and sized quantum dot semiconductor nanoparticles through a series of pH washes. This is an alternative to the apoferritin cage technique currently used to synthesize uniform CdSe nanoparticles. Such materials could also be used for targeted drug delivery since particles release contents upon exposure to specific pH levels.

Biomanufacturing

Biomanufacturing is a new type of production that uses biological systems to construct commercially-relevant biomaterials to add to medicine, industrial applications and the food and beverage industry. Biomanufactured products are found in natural sources like cultures of microbes, blood or plant and animal cells that have been artificially grown in specialized equipment. The cells used in the production may have occurred naturally or through advanced genetic engineering techniques.

Biomanufactured Products

Most consumers are unaware that there are thousands of biomanufactured products available on the market today. They are classified by medicine, food and beverage and industrial applications. In the medicine field, amino acids, vaccines, cytokines, fusion proteins, growth factors, biopharmaceuticals and monoclonal antibodies all utilize biomanufactured products. Again, amino acids are biomanufactured for the food and beverage industry to improve the health of the food or drink. Other biomanufactured products found in this category are enzymes and protein supplements.

Finally, industrial applications that require cells or enzymes that are biomanufactured include biocementation, detergents, plastics and bioremediation. Biocementation is microbiologically-induced calcium carbonate precipitation which is commonly used in the treatment of concrete to prolong service life, as a result of the added calcium, as well as in bricks and fillers for plastics and ink. Bioremediation is a technique used in waste management that involves organisms neutralizing pollutants and other contaminants. The organisms break down hazardous substances into much less toxic (or non-toxic) materials.

Methods of Biomanufacturing

While there are many methods of biomanufacturing useful cells, the following are the most popular:

Blood Plasma Fractionation

Blood plasma fractionation is the process of separating components of blood plasma which is the liquid portion of the blood. It contains numerous proteins such as immunoglobulins and albumin as well as clotting proteins like fibrinogen. These are most commonly used in medicines to improve the immune system.

Cell Culture

Cell culture is the process by which cells are grown in an artificial environment under controlled conditions. As previously mentioned, animal and plant cells are replicated and grown using this method in a culture dish. This is beneficial for medicines, supplements and enzymes.

Column Chromatography

Column chromatograph is the method used to purify chemical compounds from mixtures. This process is typically used for pharmaceuticals and medicines and offers the benefits of a relatively low cost and easily disposable.

Fermentation

Fermentation is the process known to produce alcohol. While this is a great process for a Saturday night, it also has functional purposes. Fermentation is a metabolic process converting sugar into gases, acids and alcohol. This is most important for the food and beverage industry as well as with industrial applications.

Homogenization

Homogenization is the process by which a biological sample is to a point in which all fractions are equal in composition. This is useful in medicine and industrial applications to create a balance amongst the cell constructions.

Ultrafiltration

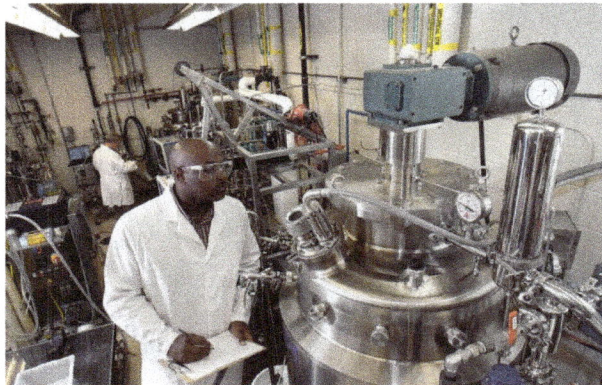

The ultrafiltration process involves the separation of solids and solutes with high molecular weights which need to be retained from the process. Low molecular weights pass through the semipermeable membrane. The process is used in drinking water, waste water treatment and other requirements for recycling processes.

Types of Equipment

The type of equipment and manufacturing facility requirements are dictated by the products being produced. The majority of process equipment is constructed from plastic or stainless steel because their chemical constructions do not interfere with process. Also, stainless steel is easy to clean and can be reused for other functions. Unfortunately some plastic equipment must be disposed of following a single use.

Those products manufactured for food and beverage or medical usage must be produced in a facility that has been designed within the confines of Good Manufacturing Practices (GMP) regulations. In many instances, cleanrooms are required to control the particulates and microorganisms. Also, aseptic processing equipment and sterilization are required for any molecules that will be part of an injectable product.

Amazing New Products

The following are the newest products from biomanufacturing:

Absorbable Heart Stent

The absorbable heart stent, created by Abbott Laboratories in Illinois, has developed a bio-absorbable version of the traditionally-metal stent for the heart. A stent opens the arteries that have become blocked due to coronary artery disease. Stents typically release medication into the artery that eliminates narrowing. A bio-absorbable version does the job and then begins to dissolve after six months. Within two years, it has disappeared from the body leaving a healthy artery.

Nerve Regenerator

When nerve fibers cannot grow along injured spinal cords due to scar tissue, an injected liquid of nanogel can create microscopic scaffolds in the nanofibers. Peptides in the fibers instruct stem cells to produce new cells thus encouraging nerve development. The scaffold supports growth of axons along the entire spinal cord. This technology was developed by Northwestern University in Illinois.

Smart Pill

the smart pill consists of sensors that track medication usage by recording the precise moment the drugs are ingested. The pill features sand-grain-size microchips that emit high frequency currents. On the skin, are Band-Aid-like receivers that log the data as well as respiration and monitor the heart rate. This information is wirelessly transmitted to a computer for analysis. Experts agree that to continuously improve pharmaceuticals, embedded digital technology must be utilized within the body and networked to receivers and computers outside of the body for thorough evaluation.

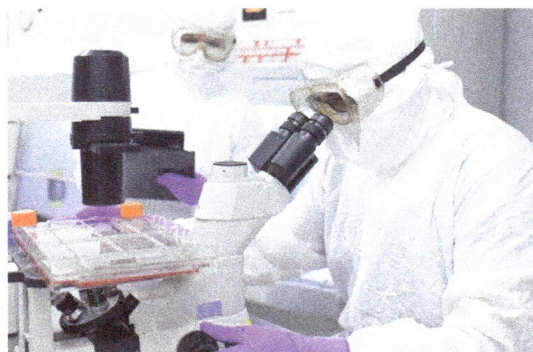

Biobased Economy

Biobased economy, bioeconomy or biotechonomy refers to all economic activity derived from scientific and research activity focused on biotechnology. In other words, understanding mechanisms and processes at the genetic and molecular levels and applying this understanding to creating or improving industrial processes.

The term is widely used by regional development agencies, international organizations, biotechnology companies. It is closely linked to the evolution of the biotechnology industry. The ability to study, understand and manipulate genetic material has been possible due to scientific breakthroughs and technological progress.

The evolution of the biotechnology industry and its application to agriculture, health, chemical or energy industries is a classic example of bioeconomic activity.

In Practice

The biobased economy uses first-generation biomass (crops), second-generation biomass (crop refuge), and third-generation biomass (seaweed, algae). Several methods of processing are then used (in biorefineries) to gather the most out of the biomass. This includes techniques such as:

- Anaerobic digestion,

- Pyrolysis,

- Torrefaction,

- Fermentation.

Anaerobic digestion is generally used to produce biogas, fermentation of sugars produces ethanol, pyrolysis is used to produce pyrolysis-oil (which is solidified biogas), and torrefaction is used to create biomass-coal. Biomass-coal and biogas is then burnt for energy production, ethanol can be used as a (vehicle)-fuel, as well as for other purposes, such as skincare products.

Getting the most out of the Biomass

For economic reasons, the processing of the biomass is done according to a specific pattern. This pattern, as well as the quantities, depends on the types of biomass used. The whole of finding the most suitable pattern is known as biorefining. A general list shows the products with high added value and lowest volume of biomass to the products with the lowest added value and highest volume of biomass:

- Fine chemicals/medicines,

- Food,

- Chemicals/bioplastics,

- Transport fuels,

- Electricity and heat.

Some research is being conducted as well in order to improve the manufacturing processes. For example, to make plastics, paint, medicines, antifreeze out of syngas, a new catalyst has been invented by Krijn de Jong.

Compared to Fossil Fuel Economy

- With a fossil fuel economy substances as gasoline, fuel oil, diesel, naphta, kerosine, LPG, and other are converted to: energy, chemical products, food, materials.

- With a biobased economy substances as (syn)gas, sugars, oil, fibres and other are converted to energy, chemical products, (animal) food, biomaterials.

References

- Genetic-engineering, science: britannica.com, Retrieved 14 July, 2019

- Isaaa 2013 annual report executive summary, global status of commercialized biotech/gm crops: 2013 isaaa brief 46-2013, retrieved 6 august 2014

- Economic impact of transgenic crops in developing countries. Agbioworld.org. Retrieved 8 february 2011

- Bashshur, ramona (february 2013). "fda and regulation of gmos". American bar association. Retrieved february 24, 2016

- Biomanufacturing: moneyinc.com, Retrieved 17 May, 2019

- "questions & answers on food from genetically engineered plants". Us food and drug administration. 22 jun 2015. Retrieved 29 september 2015

5
Fields Related to Biotechnology

Environmental biotechnology, agricultural biotechnology, industrial biotechnology, bioinformatics, biomedical engineering, nanobiotechnology, biomedicine and molecular biology are some of the fields which are closely associated with biotechnology. All these diverse fields of biotechnology have been carefully analyzed in this chapter.

Environmental Biotechnology

Environmental biotechnology in particular is the application of processes for the protection and restoration of the quality of the environment.

Environmental biotechnology can be used to detect, prevent and remediate the emission of pollutants into the environment in a number of ways.

Solid, liquid and gaseous wastes can be modified, either by recycling to make new products, or by purifying so that the end product is less harmful to the environment. Replacing chemical materials and processes with biological technologies can reduce environmental damage.

In this way environmental biotechnology can make a significant contribution to sustainable development. Environmental Biotechnology is one of today's fastest growing and most practically useful scientific fields. Research into the genetics, biochemistry and physiology of exploitable microorganisms is rapidly being translated into commercially available technologies for reversing and preventing further deterioration of the earth's environment.

Objectives of Environmental Biotechnology

The aim of environmental biotechnology is to prevent, arrest and reverse environmental degradation through the appropriate use of biotechnology in combination with other technologies, while supporting safety procedures as a primary component of the programme.

Specific Objectives are:

- To adopt production processes that make optimal use of natural resources, by recycling biomass, recovering energy and minimizing waste generation.

- To promote the use of biotechnological techniques with emphasis on bioremediation of land and water, waste treatment, soil conservation, reforestation, afforestation and land rehabilitation.

- To apply biotechnological processes and their products to protect environmental integrity with a view to long-term ecological security.

Use of biotechnology to treat pollution problems is not a new idea. Communities have depended on complex populations of naturally occurring microbes for sewage treatment for over a century. Every living organism—animals, plants, bacteria and so forth—ingests nutrients to live and produces a waste as a by-product. Different organisms need different types of nutrients.

Certain bacteria thrive on the chemical components of waste products. Some microorganisms feed on materials toxic to others. Research related environmental biotechnology is vital in developing effective solutions for mitigating, preventing and reversing environmental damage with the help of these living forms. Growing concern about public health and the deteriorating quality of the environment has prompted the development of a range of new, rapid analytical devices for the detection of hazardous compounds in air, water and land. Recombinant DNA technology has provided the possibilities for the prevention of pollution and holds a promise for a further development of bioremediation.

Applications of Environmental Biotechnology

Environmental protection is an integral component of sustainable development. The environment is threatened every day by the activities of man. With the continued increase in the use of chemicals, energy and non-renewable resources by an expanding global population, associated environmental problems are also increasing. Despite escalating efforts to prevent waste accumulation and to promote recycling, the amount of environmental damage caused by over-consumption, the quantities of waste generated and the degree of unsustainable land use appear likely to continue growing.

The remedy can be achieved, to some extent, by the application of environmental biotechnology techniques, which use living organisms in hazardous waste treatment and pollution control. Environmental biotechnology includes a broad range of applications such as bioremediation, prevention, detection and monitoring, genetic engineering for sustainable development and better quality of living.

Bioremediation

Bioremediation refers to the productive use of microorganisms to remove or detoxify pollutants, usually as contaminants of soils, water or sediments that otherwise intimidate human health. Bio treatment, bio reclamation and bio restoration are the other terminologies for bioremediation. Bioremediation is not a new practice. Microorganisms have been used for many years to remove organic matter and toxic chemicals from domestic and manufacturing waste discharge.

However, the focus in environmental biotechnology for fighting different pollution is on bioremediation. The vast majority of bioremediation applications use naturally occurring microorganisms to identify and filter toxic waste before it is introduced into the environment or to clean up existing pollution problems.

Some more advanced systems using genetically modified microorganisms are being tested in waste treatment and pollution control to remove difficult-to-degrade materials. Bioremediation can be performed in situ or in specialized reactors (ex situ). Bioremediation by microorganisms need appropriate environment for the clean up of the polluted site.

Addition of nutrients, terminal electron acceptors (O_2/NO_2), temperature, moisture to promote the growth of a particular organism may be required for the microbial activity in the polluted site. Bioremediation operations may be made either on-site or off-site, in situ or ex situ. Bioremediation has a vast potential to clean up water and soil contaminated by a variety of hazardous pollutants, domestic wastes, radioactive wastes etc.

Biological cleaning procedures make use of the fact that most organic chemicals are subjected to enzymatic attack of living organisms. The most common approach is the use of enzymes as substitute chemical catalysts. Significant reduction or complete elimination of harsh chemicals may be achieved as is observed in leather, textile processing and pulp and paper industry.

Only 1-2g of hemicellulose is substituted for 10-15 kg of chlorine to treat 1 tonne of pulp, thereby significantly reducing the chlorinated organic effluent. Environmental protection and remediation presently combine biotechnological, chemical, physical and engineering methods.

The relative importance of biotechnology is increasing as scientific knowledge and methods improve. Its lower requirements for energy and chemicals, combined with lower production of minor wastes, make it an increasingly desirable alternative to more traditional chemical and physical methods of remediation. Applications of bioremediation for maintenance of environment are several.

Waste Water and Industrial Effluents

Water pollution is a serious problem in many countries of the world. Rapid industrialisation and urbanization have generated large quantities of waste water that resulted in deterioration of surface water resources and ground water reserves. Biological, organic and inorganic pollutants contaminate the water bodies.

In many cases, these sources have been rendered unsafe for human consumption as well as for other activities such as irrigation and industrial needs. This illustrates that degraded water quality can, in effect, contribute to water scarcity as it limits its availability for both human use and the ecosystem. Treatment of the waste water before disposal is of urgent concern worldwide.

In sewage treatment plants microorganisms are used to remove the more common pollutants from waste water before it is discharged into rivers or the sea. Increasing industrial and agricultural pollution has led to a greater need for processes that remove specific pollutants such as nitrogen and phosphorus compounds, heavy metals and chlorinated compounds.

Methods include aerobic, anaerobic and physico-chemical processes in fixed-bed filters and in bioreactors in which the materials and microbes are held in suspension. Sewage and other waste waters would, if left untreated, undergo self-purification but the process requires long exposure periods. To speed up this process bioremediation measures are used.

However, five key stages are recognized in wastewater treatment:

- Preliminary treatment – grit, heavy metals and floating debris are removed.

- Primary treatment – suspended matters are removed.

- Secondary treatment – bio-oxidize organic materials by activities of aerobic and anaerobic microorganisms.

- Tertiary treatment – specific pollutants are removed (ammonia and phosphate).

- Sludge treatment – solids are removed (final stage).

Aerobic Biological Treatment

Trickling filters, rotating biological contactors or contact beds, usually consist of an inert material (rocks/ash/wood/metal) on which the microorganisms grow in the form of a complex biofilm. These have been used for more than 70 years for sewage and waste water treatment. In these processes the degradable organic matter is oxidized by the microorganisms to CO_2 that can be vented to the atmosphere.

Activated Sludge Process

This process is used for treatment and removal of dissolved and biodegradable wastes, such as organic chemicals, petroleum refining wastes textile wastes and municipal sewage. The microorganisms in activated sludge generally are composed of 70-90% organic and 10-30% inorganic matters.

The microorganisms found in this sludge are usually bacteria, fungi, protozoa and rotifers. Petroleum hydrocarbons are degraded by species of bacteria (Acinetobacter, Mycobacteria, Pseudomonas etc.), yeasts, Cladosporium and Scolecobasidium. Pesticides (aldrin, dieldrin, parathion, malathion) are detoxified by fungus Xylaria xylestrix. Pseudomonas (a predominant soil microrganism) can detoxify organic compounds like hydrocarbons, phenols, organophosphates, polychlorinated biphenyls and polycyclic aromatics.

Utilisation of immobilized cyanobacterium Phormidium laminosum in batch and continuous flow bioreactors for the removal of nitrate, nitrite and phosphate from water has been reported by Garbisu et al. Blanco et al. showed the biosorption of heavy metal by Phormidium laminosum immobilised in micro-porous polymeric matrices. Photo-bioreactors are currently used to grow algae and cyanobacteria under closely controlled environmental conditions, with a view to making high-value products (such as beta-carotene and gamma-linoleic acid), designing efficient effluent treatment processes, and providing new energy sources.

The costs of wastewater treatment can be reduced by the conversion of wastes into useful products. Sulphur metabolizing bacteria can remove heavy metals and sulphur compounds from waste streams of the galvanization industry and reused. Most anaerobic wastewater treatment systems produce useful biogas.

In some cases, the by-products of the pollution-fighting microorganisms are themselves useful.

Methane, for example, can be derived from a form of bacteria that degrades sulphur liquor, a waste product of paper manufacturing.

Soil and Land Treatment

As the human population grows, its demand for food from crops increases, making soil conservation crucial. Deforestation, over-development, and pollution from man-made chemicals are just a few of the consequences of human activity and carelessness. The increasing amounts of fertilizers and other agricultural chemicals applied to soils and industrial and domestic waste-disposal practices, led to the increasing concern of soil pollution. Pollution in soil is caused by persistent toxic compounds, chemicals, salts, radioactive materials, or disease-causing agents, which have adverse effects on plant growth and animal health.

Many species of fungi can be used for soil bioremediation. Lipomyces sp. can degrade herbicide paraquat. Rhodotorula sp. can convert benzaldehyde to benzyl alcohol. Candida sp. degrades formaldehyde in the soil. Aspergillus niger and Chaetomium cupreum are used to degrade tannins (found in tannery effluents) in the soil thereby helping in plant growth.

Phanerochaete chrysosporium has been used in bioremediation of soils polluted with different chemical compounds, usually recalcitrant and regarded as environmental pollutants. Decrease of PCP (Pentachlorophenol) between 88-91% within six weeks was observed in presence of Phanerochaete chrysosporium.

Bioremediation of contaminated soil has been used as a safe, reliable, cost-effective and environment friendly method for degradation of various pollutants. This can be effected in a number of ways, either in situ or by mechanically removing the soil for treatment elsewhere.

In situ treatments include adding nutrient solutions, introducing microorganisms and ventilation. Ex situ treatment involves excavating the soil and treating it above ground, either as compost, in soil banks, or in specialised slurry bioreactors. Bioremediation of land is often cheaper than physical methods and its products are largely harmless.

During biological treatment soil microorganisms convert organic pollutants to CO_2, water and biomass. Degradation can take place under aerobic as well as under anaerobic conditions. Soil bioremediation can also be accomplished with the help of bioreactors. Degradation can take place under aerobic as well as under anaerobic conditions. Soil bioremediation can also be accomplished with the help of bioreactors. Liquids, vapours, or solids in a slurry phase are treated in a reactor. Microbes can be of natural origin, cultivated or even genetically engineered.

Research in the field of environmental biotechnology has made it possible to treat soil contaminated with mineral oils. Solid-phase technologies are used for petroleum-contaminated soils that are excavated, placed in a containment system through which water and nutrients percolate. Biological degradation of oils has proved commercially viable both on large and small scales, in situ and ex situ.

In situ soil bioremediation involve the stimulation of indigenous microbial populations (e.g. by adding nutrients or aeration). In this process the environmental conditions for the biological degradation of organic pollutants are optimized as far as possible. Oxygen has to be supplied by artificial aeration or by adding electron acceptors such as nitrates or oxygen releasing compounds. Ozone dissolved in water and H_2O_2 are sometimes used which degrade the organic contaminants.

Air and Waste Gases

With the onset of human civilization, the air is one of the first and most polluted components of the atmosphere. Most air pollution comes from one human activity: burning fossil fuels—natural gas, coal, and oil—to power industrial processes and motor vehicles. When fuels are incompletely burned, various chemicals called volatile organic chemicals (VOCs) also enter the air. Pollutants also come from other sources.

For instance, decomposing garbage in landfills and solid waste disposal sites emits methane gas, and many household products give off VOCs. Expanding industrial activities have added more contaminants in the air.

The concept of biological air treatment at first seemed impossible. With the development of biological waste gas purification technology using bioreactors—which includes bio filters, bio trickling filters, bio scrubbers and membrane bioreactors—this problem is taken care of. The mode of operation of all these reactors is similar.

Air containing volatile compounds is passed through the bioreactors, where the volatile compounds are transferred from the gas phase into the liquid phase. Microbial community (mixture of different bacteria, fungi and protozoa) grow in this liquid phase and remove the compounds acquired from the air.

In the bio filters, the air is passed through a bed packed with organic materials that supplies the necessary nutrients for the growth of the microorganisms. This medium is kept damp by maintaining the humidity of the incoming air. Biological off-gas treatment is generally based on the absorption of the VOC in the waste gases into the aqueous phase followed by direct oxidation by a wide range of voracious bacteria, which include Nocardia sp. and Xanthomonas sp.

Prevention

Sustainable development and quality living depends upon the rational, eco-friendly use of natural resources with economic growth. To comply with this trend, industrial development has to change to sustainable style from degradative type and for such a purpose cleaner technologies have to be adopted.

According to United Nations Environment Programme 'the continuous application of an integrated preventive environmental strategy to processes, products and services to increase eco-efficiency and reduce risks to humans and the environment' defines the eco-friendly concept. The application of preventive and clean concept can only be achieved by the 5R policies.

Five Environmental Buzzwords are the 5Rs for Efficient Use of Energy and Better Control of Waste, Which Might Help in Sustainable Development and Quality Living:

- Reduce (Reduction of waste).

- Reuse (Efficient use of water, energy).

- Recycle (Recycling of wastes).

- Replace (Replacement of toxic/hazardous raw materials for more environment- friendly inputs).

- Recover (useful non-toxic fractions from wastes).

Innovation and adoption of clean technologies is the target of research and development worldwide. Industrial companies are developing processes with reduced environmental impact responding to the international call for the development of a sustainable society. There is a pervading trend towards less harmful products and processes; away from "end-of-pipe" treatment of waste streams. Environmental biotechnology, with its appropriate technologies, is suitable to contribute to this trend.

Enzyme Application

Enzymes are widely employed in industries for many years. Enzymes, non-toxic and biodegradable, are biological catalysts that are highly competent and have numerous advantages over non-biological catalysts. The use of enzyme by man, both directly and indirectly, have been for thousands of years.

In the recent years enzymes have played important roles in the production of drugs, fine chemicals, amino acids, antibiotics and steroids. Industrial processes can be made eco-friendly by the use of enzymes. Enzyme application in the textile, leather, food, pulp and paper industries help in significant reduction or complete elimination of severe chemicals and are also more economic in energy and resource consumption.

Biotechnological methods can produce food materials with improved nutritional value, functional characteristics, shelf stability. Plant cells grown in fermenters can produce flavours such as vanilla, reducing the need for extracting the compounds from vanilla beans. Food processing has benefited from biotechnologically produced chymosin which is used in cheese manufacture; alpha-amylase, which is used in production of high-fructose corn syrup and dry beer; and lactase, which is added to milk to reduce the lactose content for persons with lactose intolerance.

Genetically engineered enzymes are easier to produce than enzymes isolated from original sources and are favoured over chemically synthesized substances because they do not create by-products or off-flavours in foods.

Environmental Detection and Monitoring

A wide range of biological methods are in use to detect pollution and for the continuous monitoring of pollutants. The techniques of biotechnology have novel methods for diagnosing environmental problems and assessing normal environmental conditions so that human beings can be better- informed of the surroundings. Applications of these methods are cheaper, faster and also portable.

Rather than gathering soil samples and sending them to a laboratory for analysis, scientists can measure the level of contamination on site and know the results immediately. Biological detection methods using biosensors and immunoassays have been developed and are now in the market. Microbes are used in biosensors contamination of metals or pollutants. Saccharomyces cerevisiae (yeast) is used to detect cyanide in river water while Selenastrum capricornatum (green alga) is used for heavy metal detection. Immunoassays use labelled antibodies (complex proteins produced in biological response to specific agents) and enzymes to measure pollutant levels. If a pollutant is present, the antibody attaches itself to it making it detectable either through colour change, fluorescence or radioactivity.

Biosensors

A biosensor is an analytical device that converts a biological response into an physical, chemical or electrical signal. The development of biosensors involves integration of a specific and sensitive biologically derived sensing elements (immobilized cells, enzymes or antibodies) are integrated with physico-chemical transducers (either electrochemical or optical). Immobilised on a substrate, their properties change in response to some environmental effect in a way that is electronically or optically detectable.

It is then possible to make quantitative measurements of pollutants with extreme precision or to very high sensitivities. The biological response of the biosensor is determined by the bio catalytic membrane, which accomplishes the conversion of reactant to product. Immobilised enzymes possess a number of advantageous features which makes them particularly applicable for use in such systems.

They may be re-used, which ensures that the same catalytic activity is present for a series of analyses. Biosensors are powerful tools, which rely on biochemical reactions to detect specific substances, which have brought benefits to a wide range of sectors, including the manufacturing, engineering, chemical, water, food and beverage industries. They are able to detect even small amounts of their particular target chemicals, quickly, easily and accurately.

For this character of biosensors they have been ardently adopted for a variety of process monitoring applications, principally in respect to pollution assessment and control. Biosensors for detection of carbohydrates, organic acids, glucosinolates, aromatic hydrocarbons, pesticides, pathogenic bacteria and others have already been developed.

The biosensors can be designed to be very selective, or sensitive to a broad range of compounds. For example, a wide range of herbicides can be detected in river water using algal-based biosensors; the stresses inflicted on the organisms being measured as changes in the optical properties of the plant's chlorophyll. Biosensors are of different types such as calorimetric biosensors, immunosensors, optical biosensors, BOD biosensors, gas biosensors.

The remarkable ability of microbes to break down chemicals is proving useful, not only in pollution remediation but also in pollutant detection. A group of scientists at Los Alamos National Laboratory work with bacteria that degrade a class of organic chemicals called phenols. When the bacteria ingest phenolic compounds, the phenols attach to a receptor.

The phenol-receptor complex then binds to DNA, activating the genes involved in degrading phenol. The Los Alamos scientists added a reporter gene that, when triggered by a phenol-receptor complex, produces an easily detectable protein, thus indicating the presence of phenolic compounds in the environment. Biosensors employing acetylcholine esterase can be used for the detection of organophosphorus compounds in water.

Genetic Engineering

Biotechnology, which is expected to make a great contribution to the welfare of mankind, is an important technology that should be steadily developed. The application of DNA technology, among the different kinds of biotechnology, has the possibility to create new gene combinations that have not previously existed in nature.

Since its beginning, genetic engineering has claimed to be able to construct tailor-made microorganisms with improved degrading capabilities for toxic substances. With the development of GEM (genetically engineered microorganism) and their possible utilization in the treatment of contaminated soil and water, stability of plasmids is extremely desirable. Plasmids are circular strands of DNA that replicates as separate entities independent of the host chromosome. Plasmids can range in size from those that carry only a couple of genes to ones carrying much greater numbers. Small plasmids may be present as multiple copies. Exchange of genetic information via plasmids is achieved by the process of conjugation.

The use of restriction enzymes has enabled the isolation of particular DNA fragments that can be transferred to another organism lacking the same. Genes which code for metabolism of environmental pollutants such as PCB's and other xenobiotic compounds are frequently, although not always, located on plasmids.

The possibility of genetic transfer in non-biodegradative microbes has opened a new outlook of bio treatment of wastes. The recombinant DNA has the ability to multiply and may also confer the specific derivative capacity to detoxify environmental contaminants.

Gene transfer among microbial communities has improved the derivative capacity in vitro. The first patent for a genetically modified organism (GMO) or GEM, filed in the USA by Professor A. M. Chakrabarty was for a bacterium Pseudomonas putida with hydrocarbon degrading abilities. Subsequent reports have noted the role of plasmids in degradation of alkanes, naphthalene, toluene, m— and p— xylenes.

Given the overwhelming diversity of species, biomolecules and metabolic pathways on this planet, genetic engineering can, in principle, be a very powerful tool in creating environmentally friendlier alternatives for products and processes that presently pollute the environment or exhaust its non-renewable resources.

Nowadays organisms can be supplemented with additional genetic properties for the biodegradation of specific pollutants if naturally occurring organisms are not able to do that job properly or not quickly enough. By combining different metabolic abilities in the same microorganism blockage in environmental cleanup may be circumvented.

In the USA some genetically modified bacteria have been approved for bioremediation purposes but large scale applications have not yet been reported. In Europe only controlled field tests have been authorized. Just as light, heat, and moisture can degrade many materials, biotechnology relies on naturally occurring, living bacteria to perform a similar function but the action is faster.

Some bacteria naturally feed on chemicals and other wastes, including some hazardous materials. They consume those materials, digest them, and excrete harmless substances in their place. Bioremediation uses natural as well as recombinant microorganisms to break down toxic and hazardous substances already present in the environment. Bio treatment can be used to detoxify waste streams at the source before they contaminate the environment – rather than at the point of disposal. This approach involves carefully selecting organisms, known as biocatalysts, which are enzymes that degrade specific compounds and accelerate the degradation process.

However, the application of GMOs/GEMs, in the environment for bioremediation may create problems in the ecosystem. These exclusively designed organisms do not get a chance to experience the various fluctuating environmental conditions which is faced by naturally occurring organisms during the evolutionary processes spaning millions of years.

As a result, the latter are well adapted to the changing environmental conditions such as changes in temperature, substrate or waste concentrations. But when exposed to the contaminated site, GMOs show a higher viability than naturally occurring bacteria, due to their tailored enzymatic equipment.

There are concerns about the negative effect of these GMOs on the complex and delicate microbial ecosystems by competition or the exchange of genetic material in the soils to which they are applied. Even more worrisome is their potential effect outside the treatment area. While recombinant strains may appear harmless in the laboratory, it is virtually impossible to assess their impact in the field.

Biotechnical methods are now used to produce many proteins for pharmaceutical and other specialized purposes. Human insulin, the first genetically engineered product to be produced commercially (1982) is made by nonvirulent strain of Escherichia coli bacteria, by introduction of a copy of the gene for human insulin.

When the gene is "amplified" the bacterial cells produce large quantities of human insulin that are purified and used to treat diabetes in human beings. A number of other genetically engineered products have been approved since then, including human growth hormone, alpha interferon, recombinant erythropoietin and tissue plasminogen activator.

Biotechnology techniques are being applied to plants to produce plant materials with improved composition, functional characteristics. Among the first commercially available whole food products was the slow-ripening tomato, the gene for polygalacturonase, the enzyme responsible for softening, is turned off in this tomato. Plants that are resistant to disease, pests, environmental conditions, or selected herbicides or pesticides are also being developed.

In 1995, the Environmental Protection Agency (EPA) gave clearance for development of transgenic corn seed, cotton seed, and seed potatoes that contain the genetic material to resist certain insects. The advantage of such products is that they allow the use of less toxic and more environmentally friendly herbicides and pesticides.

The first approved application of biotechnology to animal production was the use of recombinant bovine somatotropin (BST) in dairy cows. Bovine somatotropin, a protein hormone found naturally in cows, is necessary for milk production. When the recombinant BST is administered to dairy cows under ideal management conditions, milk production has been shown to increase by 10% to 25%.

Other uses of biotechnology in animal production include development of vaccines to protect animals from disease, production of several calves from one embryo (cloning), artificial insemination, improvement of growth rate and/or feed efficiency, and rapid disease detection.

Natural bio-pesticides are another development of biotechnology that help farmers reduce chemical use. They degrade rapidly, leave no residues, and are toxic only to target insects. Bacillus thuringiensis (B.t.), produces a protein that is naturally toxic to certain insects. Scientists have extracted the B.t. gene that expresses the insecticide and inserted it into common bacteria that can be grown in large quantities by the same fermentation techniques used to produce such everyday products as beer and antibiotics. Spread on cotton and other crops, these harmless bacteria control insects naturally.

Moreover, a wide range of crop plants have been genetically engineered to express the cry genes (found in B. t.) in their tissues, so the insects get killed as they feed on these crops. Pollution control by genetic engineering is likely to work best when pollutants are a known mixture of relatively concentrated organic compounds that are related to each other in structure, where conventional alternative organic nutrients are absent, and when there is no competition from indigenous microorganisms.

The spectacular metabolic versatility of bacteria and fungi is exploited in the area environmental bioremediation as in sewage and waste water treatment, degradation of xenobiotics and metal abatement. Genetic manipulation offers a way of engineering microorganisms to deal with a pollutant, or a family of closely related pollutants, that may be present in the waste stream from an industrial process.

The simplest approach is to extend the degradative capabilities of existing metabolic pathways within an organism either by introducing additional enzymes from other organisms or by modifying the specificity or catalytic mechanisms of enzymes already present.

A treatment plant at the Homestake Mine in Lead, South Dakota, purifies 4 million gallons of cyanide-containing wastewater a day by completely converting cyanide to nitrate. Pseudomonas sp. convert cyanide and thiocyanate to ammonia and bicarbonate and the nitrifying bacteria Nitrosomonas and Nitrobacter cooperate in converting ammonia to nitrate. Recombinant DNA technology has had amazing repercussion in the last few years in environmental protection and also in other fields for better quality of living.

Different Areas of Environmental Biotechnology

Environmental Biotechnology and Metagenomics

Environmental Biotechnology is Divided into Different Areas:

- Direct studies of the environment,

- Research with a focus on applications to the environment,

- Research that applies information from the environment to other venues.

Here, a brief account of a particular aspect of direct analysis of environment is given.

In addition to DNA inside living organisms, there is much free DNA in the environment that might also be a source of new genes. The field of environmental biotechnology has revolutionized the study of the life-forms which have not been studied earlier and DNA.

This approach is direct analyses of the environment and the natural biochemical processes that are present. A significant study in this aspect is metagenomics. Metagenomics is the study of the genomes of whole communities of microscopic life forms and it deals with a mixture of DNA from multiple organisms, viruses, viroids, plasmids and free DNA.

In other words, metagenomics, the genomic analysis of a population of microorganisms, is the method to gain access to the physiology and genetics of uncultured organisms.

Using metagenomics, researchers investigate, catalogue the current microbial diversity. New proteins, enzymes and biochemical pathways are identified. The knowledge garnered from metagenomics has the potential to affect the ways we use the environment. Metagenomic analyses involves isolating DNA from an environmental sample, cloning the DNA into a suitable vector, transforming the clones into a host bacterium and screening the resultant transformants.

The clones can be screened for phylogenetic markers such as 16S rRNA and rec A or for other conserved genes by hybridization or multiplex PCR or for expression of specific traits such as enzyme activity or antibiotic production or they can be sequenced randomly.

One very important method for metagenomic study is stable isotope probing (SIP). An environmental sample of water or soil is first mixed with a precursor such as methanol, phenol, carbonate or ammonia that has been labeled with a stable isotope such as ^{15}N, ^{13}C or ^{18}O. If the organisms in the sample metabolize the precursor substrate, the stable isotope is incorporated into their genome.

When the DNA from the sample is isolated and separated by centrifugation, the genomes that incorporated the labeled substrate will be heavier and can be separated from the other DNA in the sample. The heavier DNA will migrate further in a cesium chloride gradient during centrifugation. The DNA can be used directly or cloned into vectors to make a metagenomic library. This technique is useful to find new organisms that can degrade contaminants such as phenol.

Microorganisms are crucial participants in cleaning up a large variety of hazardous substances/chemicals by transforming them into forms that are harmless to people and environment. One very important example is given here. Gasoline is leaked into soil in every gas station in United States.

There is every possibility that gasoline will be mixed with ground water which is the prime source of drinking water. However, the dormant members of the soil microbial community are triggered to become active and degrade the harmful chemicals in gasoline.

Since gasoline is composed of hundreds of chemicals it takes a variety of microbes working together to degrade them all. When some bacteria cause a depletion of O_2 in ground water near a gasoline spill, other types of bacteria that can use nitrate for energy begin biodegrading the gasoline. Bacteria that use iron, manganese and sulfate follow.

All these microbial communities work together in a pattern to transform leaking gasoline into CO_2 and water. Metagenomic analysis may help us identify the particular community member and function needed to achieve the full chemical transformation that will keep our planet livable.

Agricultural Biotechnology

Agricultural biotechnology, also known as agritech, is an area of agricultural science involving the use of scientific tools and techniques, including genetic engineering, molecular markers, molecular diagnostics, vaccines, and tissue culture, to modify living organisms: plants, animals, and micro-organisms. Crop biotechnology is one aspect of agricultural biotechnology which has been greatly developed upon in recent times. Desired trait are exported from a particular species of Crop to an entirely different species. These transgene crops possess desirable characteristics in terms of flavor, color of flowers, growth rate, size of harvested products and resistance to diseases and pests.

Crop Modification Techniques

Traditional Breeding

Traditional crossbreeding has been used for centuries to improve crop quality and quantity. Crossbreeding mates two sexually compatible species to create a new variety with the desired traits of the parents. For example, the honeycrisp apple exhibits a specific texture and flavor due to the crossbreeding of its parents. In traditional practices, pollen from one plant is placed on the female part of another, which leads to a hybrid that contains genetic information from both parent plants. Plant breeders select the plants with the traits they're looking to pass on and continue to breed those plants. Note that crossbreeding can only be utilized within the same or closely related species.

Mutagenesis

Mutations can occur randomly in the DNA of any organism. In order to create variety within crops, scientists can randomly induce mutations within plants. Mutagenesis uses radioactivity to induce random mutations in the hopes of stumbling upon the desired trait. Scientists can use mutating chemicals such as ethyl methanesulfonate, or radioactivity to create random mutations within the DNA. Atomic gardens are used to mutate crops. A radioactive core is located in the center of a circular garden and raised out of the ground to radiate the surrounding crops, generating mutations within a certain radius. Mutagenesis through radiation was the process used to produce ruby red grapefruits.

Polyploidy

Polyploidy can be induced to modify the number of chromosomes in a crop in order to influence its fertility or size. Usually, organisms have two sets of chromosomes, otherwise known as a diploidy. However, either naturally or through the use of chemicals, that number of chromosomes can change, resulting in fertility changes or size modification within the crop. Seedless watermelons are created in this manner; a 4-set chromosome watermelon is crossed with a 2-set chromosome watermelon to create a sterile (seedless) watermelon with three sets of chromosomes.

Protoplast Fusion

Protoplast fusion is the joining of cells or cell components to transfer traits between species. For example, the trait of male sterility is transferred from radishes to red cabbages by protoplast fusion. This male sterility helps plant breeders make hybrid crops.

RNA Interference

RNA interference (RNAIi) is the process in which a cell's RNA to protein mechanism is turned down or off in order to suppress genes. This method of genetic modification works by interfering with messenger RNA to stop the synthesis of proteins, effectively silencing a gene.

Transgenics

Transgenics involves the insertion of one piece of DNA into another organism's DNA in order to introduce a new gene(s) into the original organism. This addition of genes into an organism's genetic material creates a new variety with desired traits. The DNA must be prepared and packaged in a test tube and then inserted into the new organism. New genetic information can be inserted with biolistics. An example of transgenics is the rainbow papaya, which is modified with a gene that gives it resistance to the papaya ringspot virus.

Genome Editing

Genome editing is the use of an enzyme system to modify the DNA directly within the cell. Genome editing was used to develop herbicide resistant canola to help farmers control weeds.

Improved Nutritional Content

Agricultural biotechnology has been used to improve the nutritional content of a variety of crops in an effort to meet the needs of an increasing population. Genetic engineering can produce crops with a higher concentration of vitamins. For example, golden rice contains three genes that allow plants to produce compounds that are converted to vitamin A in the human body. This nutritionally improved rice is designed to combat the world's leading cause of blindness—vitamin A deficiency. Similarly, the Banana 21 project has worked to improve the nutrition in bananas to combat micronutrient deficiencies in Uganda. By genetically modifying bananas to contain vitamin A and iron, Banana 21 has helped foster a solution to micronutrient deficiencies through the vessel of a staple food and major starch source in Africa. Additionally, crops can be engineered to reduce toxicity or to produce varieties with removed allergens.

Genes and Traits of Interest for Crops

Agronomic Traits

1. Insect Resistance:

One highly sought after trait is insect resistance. This trait increases a crop's resistance to bugs and allows for a higher yield. These genetically engineered crops can now produce their own Bt (*Bacillus thuringiensis*), which contains toxin-producing proteins that are non-harmful to humans. Bt corn and cotton are now commonplace, and cowpeas, sunflower, soybeans, tomatoes, tobacco, walnut, sugar cane, and rice are all being studied in relation to Bt.

2. Herbicide Tolerance:

Weeds have proven to be an issue for farmers for thousands of years; they compete for soil nutrients, water, and sunlight and prove deadly to crops. Biotechnology has offered a solution in the form of herbicide tolerance. Chemical herbicides are sprayed directly on plants in order to kill weeds and therefore competition, and herbicide resistant crops have to the opportunity to flourish.

3. Disease Resistance:

Often, crops are afflicted by disease spread through insects (like aphids). Spreading disease among crop plants is incredibly difficult to control and was previously only managed by completely removing the affected crop. The field of agricultural biotechnology offers a solution through genetically engineering virus resistance. Developing GE disease-resistant crops now include cassava, maize, and sweet potato.

4. Temperature Tolerance:

Agricultural biotechnology can also provide a solution for plants in extreme temperature conditions. In order to maximize yield and prevent crop death, genes can be engineered that help to regulate cold and heat tolerance. For example, papaya trees have been genetically modified in order to be more tolerant of hot and cold conditions. Other traits include water use efficiency, nitrogen use efficiency and salt tolerance.

Quality Traits

Quality traits include increased nutritional or dietary value, improved food processing and storage, or the elimination of toxins and allergens in crop plants.

Common GMO Crops

Currently, only a small number of genetically modified crops are available for purchase and consumption in the United States. The USDA has approved soybeans, corn, canola, sugar beets, papaya, squash, alfalfa, cotton, apples, and potatoes. GMO apples (arctic apples) are non-browning apples and eliminate the need for anti-browning treatments, reduce food waste, and bring out flavor. The production of Bt cotton has skyrocketed in India, with 10 million hectares planted for the first time in 2011, resulting in a 50% insecticide application reduction. In 2014, Indian and Chinese farmers planted more than 15 million hectares of Bt cotton.

Safety Testing and Government Regulations

Agricultural biotechnology regulation in the US falls under three main government agencies: The Department of Agriculture (USDA), the Environmental Protection Agency (EPA), and the Food and Drug Administration (FDA). The USDA must approve the release of any new GMOs, EPA controls the regulation of insecticide, and the FDA evaluates the safety of a particular crop sent to market. On average, it takes nearly 13 years and $130 million of research and development for a genetically modified organism to come to market. The regulation process takes up to 8 years in the United States. The safety of GMOs has become a topic of debate worldwide, but scientific articles are being conducted to test the safety of consuming GMOs in addition to the FDA's work. In one such article, it was concluded that Bt rice did not adversely affect digestion and did not induce horizontal gene transfer.

Industrial Biotechnology

Industrial biotechnology is one of the most promising new approaches to pollution prevention, resource conservation, and cost reduction. It is often referred to as the third wave in biotechnology. If developed to its full potential, industrial biotechnology may have a larger impact on the world than health care and agricultural biotechnology. It offers businesses a way to reduce costs and create new markets while protecting the environment. Also, since many of its products do not require the lengthy review times that drug products must undergo, it's a quicker, easier pathway to the market. Today, new industrial processes can be taken from lab study to commercial application in two to five years, compared to up to a decade for drugs.

The application of biotechnology to industrial processes is not only transforming how we manufacture products but is also providing us with new products that could not even be imagined a few years ago. Because industrial biotechnology is so new, its benefits are still not well known or understood by industry, policymakers, or consumers.

From the beginning, industrial biotechnology has integrated product improvements with pollution prevention. Nothing illustrates this better than the way industrial biotechnology solved the phosphate water pollution problems in the 1970s caused by the use of phosphates in laundry detergent. Biotechnology companies developed enzymes that removed stains from clothing better than phosphates, thus enabling replacement of a polluting material with a non-polluting biobased additive while improving the performance of the end product. This innovation dramatically reduced phosphate-related algal blooms in surface waters around the globe, and simultaneously enabled consumers to get their clothes cleaner with lower wash water temperatures and concomitant energy savings.

Rudimentary industrial biotechnology actually dates back to at least 6000 B.C. when Neolithic cultures fermented grapes to make wine, and Babylonians used microbial yeasts to make beer. Over time, mankind's knowledge of fermentation increased, enabling the production of cheese, yogurt, vinegar, and other food products. In the 1800s, Louis Pasteur proved that fermentation was the result of microbial activity. Then in 1928, Sir Alexander Fleming extracted penicillin from mold. In the 1940s, large-scale fermentation techniques were developed to make industrial quantities of this wonder drug. Not until after World War II, however, did the biotechnology revolution begin, giving rise to modern industrial biotechnology.

Since that time, industrial biotechnology has produced enzymes for use in our daily lives and for the manufacturing sector. For instance, meat tenderizer is an enzyme and some contact lens cleaning fluids contain enzymes to remove sticky protein deposits. In the main, industrial biotechnology involves the microbial production of enzymes, which are specialized proteins. These enzymes have evolved in nature to be super-performing biocatalysts that facilitate and speed-up complex biochemical reactions. These amazing enzyme catalysts are what make industrial biotechnology such a powerful new technology.

Industrial biotechnology involves working with nature to maximize and optimize existing biochemical pathways that can be used in manufacturing. The industrial biotechnology revolution rides on a series of related developments in three fields of study of detailed information derived from the cell: genomics, proteomics, and bioinformatics. As a result, scientists can apply new techniques to a large number of microorganisms ranging from bacteria, yeasts, and fungi to marine diatoms and protozoa.

Industrial biotechnology companies use many specialized techniques to find and improve nature's enzymes. Information from genomic studies on microorganisms is helping researchers capitalize on the wealth of genetic diversity in microbial populations. Researchers first search for enzyme-producing microorganisms in the natural environment and then use DNA probes to search at the molecular level for genes that produce enzymes with specific biocatalytic capabilities. Once isolated, such enzymes can be identified and characterized for their ability to function in specific industrial processes. If necessary, they can be improved with biotechnology techniques.

Many biocatalytic tools are rapidly becoming available for industrial applications because of the recent and dramatic advances in biotechnology techniques. In many cases, the biocatalysts or whole-cell processes are so new that many chemical engineers and product development specialists in the private sector are not yet aware that they are available for deployment. This is a good example of a "technology gap" where there is a lag between availability and widespread use of a new technology. This gap must be overcome to accelerate progress in developing more economic and sustainable manufacturing processes through the integration of biotechnology.

Nanobiotechnology

Nanobiotechnology is the application of nanotechnologies in biological fields. Chemists, physicists and biologists each view nanotechnology as a branch of their own subject, and collaborations in which they each contribute equally are common. One result is the hybrid field of nanobiotechnology that uses biological starting materials, biological design principles or has biological or medical applications.

While biotechnology deals with metabolic and other physiological processes of biological subjects including microorganisms, in combination with nanotechnology, nanobiotechnology can play a vital role in developing and implementing many useful tools in the study of life.

Although the integration of nanomaterials with biology has led to the development of diagnostic devices, contrast agents, analytical tools, therapy, and drug-delivery vehicles, bionanotechnology research is still in its infancy.

Applications of Nanobiotechnology in Medicine

Nanotechnology in medicine is a wide area that encompasses disease diagnosis, target-specific drug delivery, and molecular imaging.

In particular thanks to nanoelectronics will the medical sector undergo deep changes by exploiting the traditional strengths of the semiconductor industry – miniaturization and integration. While conventional electronics have already found many applications in biomedicine – medical monitoring of vital signals, biophysical studies of excitable tissues, implantable electrodes for brain stimulation, pacemakers, and limb stimulation – the use of nanomaterials and nanoscale applications will bring a further push towards implanted electronics in the human body.

A lot of nanotechnology work is going on in the area of brain research. For instance the use of a carbon nanotube rope to electrically stimulate neural stem cells; nanotechnology to repair the brain and other advances in fabricating nanomaterial-neural interfaces for signal generation.

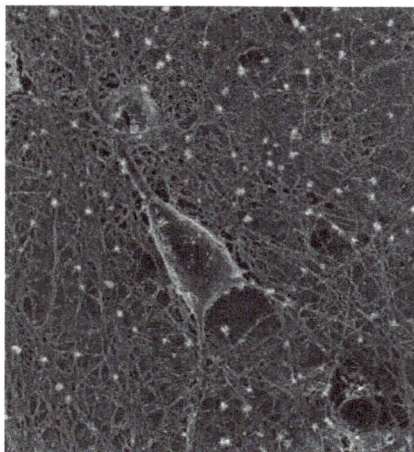

TSEM micrograph of a cultured rat hippocampal neuron grown on a layer of purified carbon nanotubes.

Sensors and Diagnostics

Molecular sensing and molecular electronics is a diverse area that can include molecular conformational changes, changes in charge distribution, changes in optical absorbance and emission, or changes in electrical conductivity along or across simple or complex-shaped molecules, all in response to a target input. Each of these approaches can be integrated into a transduction system that provides a measurable and desired change in response to a specific or range of inputs. The ability to integrate such transduction mechanisms with biomolecules or to use biomolecules as the source of such materials provides, to varying extent, biocompatibility with other systems.

Plasmonic nanobiosensors, could ultimately become a key asset in personalized medicine by helping to diagnose diseases at an early stage.

In other work, nanobiosensor that were originally designed to detect herbicides can help diagnose multiple sclerosis and a new type of smartphone-based nanobiosensor has shown promise for early detection of tuberculosis.

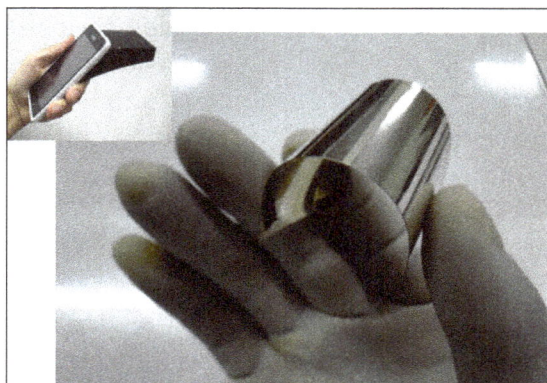

Silver and fullerene coated cellulose-acetate sensors are flexible and could be used to detect tuberculosis in early stages. The inset shows a smartphone-based detection scheme for use in point-of-care settings.

Quantum dots and noble metal nanoclusters are very active and exciting areas in the field of bionanotechnology, with new progress constantly being made in adapting these technologies in the creation of new biosensors and bioelectronic devices.

Integrating DNA and other Nucleic Acids with Nanoparticles

Functionally integrating DNA and other nucleic acids with nanoparticles in all their different physicochemical forms has produced a rich variety of composite nanomaterials which, in many cases, display unique or augmented properties due to the synergistic activity of both components. For instance, researchers have fabricated a DNA impedance biosensor for the early detection of cancer or the rapid detection of the flu virus.

These capabilities, in turn, are attracting greater attention from various research communities in search of new nanoscale tools for diverse applications that include (bio)sensing, labeling, targeted imaging, cellular delivery, diagnostics, therapeutics, theranostics, bioelectronics, and biocomputing to name just a few amongst many others.

The role of Nanobiotechnology in the Food Sector

The development of nanotechnology in food and agriculture has led to nanobiotechnology applications in that include pesticide delivery systems through bioactive nanoencapsulation; biosensors to detect and quantify pathogens; organic compounds; other chemicals and food composition alteration; high-performance sensors (electronic tongue and nose); and edible thin films to preserve fruit.

Bioinformatics

Bioinformatics is an interdisciplinary field that develops methods and software tools for understanding biological data. As an interdisciplinary field of science, bioinformatics combines biology, computer science, information engineering, mathematics and statistics to analyze and interpret biological data. Bioinformatics has been used for *in silico* analyses of biological queries using mathematical and statistical techniques.

Bioinformatics is both an umbrella term for the body of biological studies that use computer programming as part of their methodology, as well as a reference to specific analysis "pipelines" that are repeatedly used, particularly in the field of genomics. Common uses of bioinformatics include the identification of candidates genes and single nucleotide polymorphisms (SNPs). Often, such identification is made with the aim of better understanding the genetic basis of disease, unique adaptations, desirable properties (esp. in agricultural species), or differences between populations. In a less formal way, bioinformatics also tries to understand the organisational principles within nucleic acid and protein sequences, called proteomics.

Bioinformatics has become an important part of many areas of biology. In experimental molecular biology, bioinformatics techniques such as image and signal processing allow extraction of useful results from large amounts of raw data. In the field of genetics, it aids in sequencing and annotating genomes and their observed mutations. It plays a role in the text mining of biological literature and the development of biological and gene ontologies to organize and query biological data. It also plays a role in the analysis of gene and protein expression and regulation. Bioinformatics tools aid in the comparison of genetic and genomic data and more generally in the understanding of evolutionary aspects of molecular biology. At a more integrative level, it helps analyze and catalogue the biological pathways and networks that are an important part of systems biology. In structural biology, it aids in the simulation and modeling of DNA, RNA, proteins as well as biomolecular interactions.

Sequences

5'ATGACGTGGGGA3'
3'TACTGCACCCCT5'

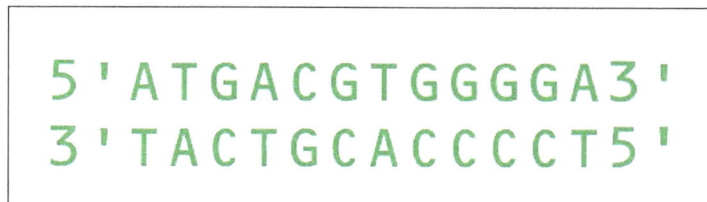

Sequences of genetic material are frequently used in bioinformatics and are easier to manage using computers than manually.

Computers became essential in molecular biology when protein sequences became available after Frederick Sanger determined the sequence of insulin in the early 1950s. Comparing multiple sequences manually turned out to be impractical. A pioneer in the field was Margaret Oakley Dayhoff. She compiled one of the first protein sequence databases, initially published as books and pioneered methods of sequence alignment and molecular evolution. Another early contributor to bioinformatics was Elvin A. Kabat, who pioneered biological sequence analysis in 1970 with his comprehensive volumes of antibody sequences released with Tai Te Wu between 1980 and 1991.

Goals

To study how normal cellular activities are altered in different disease states, the biological data must be combined to form a comprehensive picture of these activities. Therefore, the field of bioinformatics has evolved such that the most pressing task now involves the analysis and interpretation of various types of data. This includes nucleotide and amino acid sequences, protein domains, and protein structures. The actual process of analyzing and interpreting data is referred to as computational biology. Important sub-disciplines within bioinformatics and computational biology include:

- Development and implementation of computer programs that enable efficient access to, use and management of, various types of information.

- Development of new algorithms (mathematical formulas) and statistical measures that assess relationships among members of large data sets. For example, there are methods to locate a gene within a sequence, to predict protein structure and/or function, and to cluster protein sequences into families of related sequences.

The primary goal of bioinformatics is to increase the understanding of biological processes. What sets it apart from other approaches, however, is its focus on developing and applying computationally intensive techniques to achieve this goal. Examples include: pattern recognition, data mining, machine learning algorithms, and visualization. Major research efforts in the field include sequence alignment, gene finding, genome assembly, drug design, drug discovery, protein structure alignment, protein structure prediction, prediction of gene expression and protein–protein interactions, genome-wide association studies, the modeling of evolution and cell division/mitosis.

Bioinformatics now entails the creation and advancement of databases, algorithms, computational and statistical techniques, and theory to solve formal and practical problems arising from the management and analysis of biological data.

Over the past few decades, rapid developments in genomic and other molecular research technologies and developments in information technologies have combined to produce a tremendous amount of information related to molecular biology. Bioinformatics is the name given to these mathematical and computing approaches used to glean understanding of biological processes.

Common activities in bioinformatics include mapping and analyzing DNA and protein sequences, aligning DNA and protein sequences to compare them, and creating and viewing 3-D models of protein structures.

Relation to other Fields

Bioinformatics is a science field that is similar to but distinct from biological computation, while it is often considered synonymous to computational biology. Biological computation uses bioengineering and biology to build biological computers, whereas bioinformatics uses computation to better understand biology. Bioinformatics and computational biology involve the analysis of biological data, particularly DNA, RNA, and protein sequences. The field of bioinformatics experienced explosive growth starting in the mid-1990s, driven largely by the Human Genome Project and by rapid advances in DNA sequencing technology.

Analyzing biological data to produce meaningful information involves writing and running software programs that use algorithms from graph theory, artificial intelligence, soft computing, data mining, image processing, and computer simulation. The algorithms in turn depend on theoretical foundations such as discrete mathematics, control theory, system theory, information theory, and statistics.

Sequence Analysis

```
A5ASC3.1   14  SIKLWPPSQTTRLLLVERMANNLST..PSIFTRK..YGSLSKEEARENAKQIEEVACSTANQ.....HYEKEPDGDGGSAVQLYAKECSKLILEVLK 101
B4F917.1   13  SIKLWPPSESTRIMLVDRMTNNLST..ESIFSRK..YRLLGKQEAHENAKTIEELCFALADE.....HFREEPDGDGSSAVQLYAKETSKMMLEVLK 100
A9S1V2.1   23  VFKLWPPSQGTREAVRQKMALKLSS..ACFESQS..FARIELADAQEHARAIEEVAFGAAQE.....ADSGGDKTGSAVVMVYAKHASKLMLETLR 109
B9GSN7.1   13  SVKLWPPGQSTRLMLVERMTKNFIT..PSFISRK..YGLLSKEEAEEDAKKIEEVAFAAANQ.....HYEKQPDGDGSSAVQIYAKESSRLMLEVLK 100
Q8H056.1   30  SFSIWPPTQRTRDAVVRRLVDTLGG..DTILCKR..YGAVPAADAEPAARGIEAEAFDAAAA..SGEAAATASVEEGIKALQLYSKEVSRRLLDFVK 120
Q0D4Z3.2   44  SLSIWPPSQRTRDAVVRRLVQTLVA..PSILSQR..YGAVPEAEAGRAAAAVEAEAYAAVTES.SSAAAAPASVEDGIEVLQAYSKEVSRRLLELAK 135
B9MVW8.1   56  SFSIWPPTQRTRDAIISRLIETLST..TSVLSKR..YGTIPKEEASEASRRIEEEAFSGAST.......VASSEKDGLEVLQLYSKEISKRMLETVK 141
Q0IYC5.1   29  SFAVWPPTRRTRDAVVRRLVAVLSGDTTTALRKKRYRYGAVPAADAERAARAVEAQAFDAASA....SSSSSSSVEDGIETLQLYSREVSNRLLAFVR 121
A9NWJ46.1  13  SIKLWPPSESTRLMLVERMTDNLSS..VSFFSRK..YGLLSKEEAAENAKRIEETAFLAAND.....HEAKEPNLDDSSVVQFYAREASKLMLEALK 100
Q9C500.1   57  SLRIWPPTQKTRDAVLNRLIETLST..ESILSKR..YGTLKSDDATTVAKLIEEEAYGVASN.......AVSSDDDGIKILELYSKEISKRMLESVK 142
Q2HRI7.1   25  NYSIWPPKQRTRDAVKNRLIETLST..PSVLTKR..YGTMSADEASAAAIQIEDEAFSVANA.......SSSTSNDNVTILEVYSKEISKRMIETVK 110
Q9M7N3.1   28  SFKIWPPTQRTREAVVRRLVETLTS..QSVLSKR..YGVIPEEDATSAARIIEEEAFSVASV.ASAASTGGRPEDEWIEVLHIYSQEIXQRVVESAK 119
Q9M7N6.1   25  SFSIWPPTQRTRDAVINRLIESLST..PSILSKR..YGTLPQDEASETARLIEEEAFAAAGS.......TASDADDGIEILQVYSKEISKRMIDTVK 110
Q9LE82.1   14  SVKMWPPSKSTRLMLVERMTKNITT..PSIFSRK..YGLLSVEEAEQDAKRIEDLAFATANK.....HFQNEPDGDGTSAVHVYAKESSKLMLDVIK 101
Q9M651.2   13  SIKLWPPSLPTRKALIERITNNFSS..KTIFTEK..YGSLTKDQATENAKRIEDIAFSTANQ.....QFEREPDGDGGSAVQLYAKECSKLILEVLK 100
B9R748.1   48  SLSIWPPTQRTRDAVITRLIETLSS..PSVLSKR..YGTISHDEAESAARRIEDEAFGVANT.......ATSAEDDGLEILQLYSKEISRRMLDTVK 133
```

The sequences of different genes or proteins may be aligned side-by-side to measure their similarity. This alignment compares protein sequences and genomic sequences containing WPP domains.

Since the Phage Φ-X174 was sequenced in 1977, the DNA sequences of thousands of organisms have been decoded and stored in databases. This sequence information is analyzed to determine genes that encode proteins, RNA genes, regulatory sequences, structural motifs, and repetitive sequences. A comparison of genes within a species or between different species can show similarities between protein functions, or relations between species (the use of molecular systematics to construct phylogenetic trees). With the growing amount of data, it long ago became impractical to analyze DNA sequences manually. Today, computer programs such as BLAST are used daily to search sequences from more than 260 000 organisms, containing over 190 billion nucleotides. These programs can compensate for mutations (exchanged, deleted or inserted bases) in the DNA sequence, to identify sequences that are related, but not identical. A variant of this sequence alignment is used in the sequencing process itself. For the special task of taxonomic classification of sequence snippets, modern k-mer based software like Kraken achieves throughput unreachable by alignment methods.

DNA Sequencing

Before sequences can be analyzed they have to be obtained. DNA sequencing is still a non-trivial problem as the raw data may be noisy or afflicted by weak signals. Algorithms have been developed for base calling for the various experimental approaches to DNA sequencing.

Sequence Assembly

Most DNA sequencing techniques produce short fragments of sequence that need to be assembled to obtain complete gene or genome sequences. The so-called shotgun sequencing technique (which was used, for example, by The Institute for Genomic Research (TIGR) to sequence the first bacterial genome, *Haemophilus influenzae*) generates the sequences of many thousands of small DNA fragments (ranging from 35 to 900 nucleotides long, depending on the sequencing technology). The ends of these fragments overlap and, when aligned properly by a genome assembly program, can be used to reconstruct the complete genome. Shotgun sequencing yields sequence data quickly, but the task of assembling the fragments can be quite complicated for larger genomes. For a genome as large as the human genome, it may take many days of CPU time on large-memory, multiprocessor computers to assemble the fragments, and the resulting assembly usually contains numerous gaps that must be filled in later. Shotgun sequencing is the method of choice for virtually all genomes sequenced today, and genome assembly algorithms are a critical area of bioinformatics research.

Genome Annotation

In the context of genomics, annotation is the process of marking the genes and other biological features in a DNA sequence. This process needs to be automated because most genomes are too large to annotate by hand, not to mention the desire to annotate as many genomes as possible, as the rate of sequencing has ceased to pose a bottleneck. Annotation is made possible by the fact that genes have recognisable start and stop regions, although the exact sequence found in these regions can vary between genes.

The first description of a comprehensive genome annotation system was published in 1995 by the team at The Institute for Genomic Research that performed the first complete sequencing and analysis of the genome of a free-living organism, the bacterium *Haemophilus influenzae*. Owen White designed and built a software system to identify the genes encoding all proteins, transfer RNAs, ribosomal RNAs (and other sites) and to make initial functional assignments. Most current genome annotation systems work similarly, but the programs available for analysis of genomic DNA, such as the GeneMark program trained and used to find protein-coding genes in *Haemophilus influenzae*, are constantly changing and improving.

Following the goals that the Human Genome Project left to achieve after its closure in 2003, a new project developed by the National Human Genome Research Institute in the U.S appeared. The so-called ENCODE project is a collaborative data collection of the functional elements of the human genome that uses next-generation DNA-sequencing technologies and genomic tiling arrays, technologies able to automatically generate large amounts of data at a dramatically reduced per-base cost but with the same accuracy (base call error) and fidelity (assembly error).

Computational Evolutionary Biology

Evolutionary biology is the study of the origin and descent of species, as well as their change over time. Informatics has assisted evolutionary biologists by enabling researchers to:

- Trace the evolution of a large number of organisms by measuring changes in their DNA, rather than through physical taxonomy or physiological observations alone,

- Compare entire genomes, which permits the study of more complex evolutionary events, such as gene duplication, horizontal gene transfer, and the prediction of factors important in bacterial speciation,

- Build complex computational population genetics models to predict the outcome of the system over time,

- Track and share information on an increasingly large number of species and organisms.

Future work endeavours to reconstruct the now more complex tree of life.

The area of research within computer science that uses genetic algorithms is sometimes confused with computational evolutionary biology, but the two areas are not necessarily related.

Comparative Genomics

The core of comparative genome analysis is the establishment of the correspondence between genes (orthology analysis) or other genomic features in different organisms. It is these intergenomic maps that make it possible to trace the evolutionary processes responsible for the divergence of two genomes. A multitude of evolutionary events acting at various organizational levels shape genome evolution. At the lowest level, point mutations affect individual nucleotides. At a higher level, large chromosomal segments undergo duplication, lateral transfer, inversion, transposition, deletion and insertion. Ultimately, whole genomes are involved in processes of hybridization, polyploidization and endosymbiosis, often leading to rapid speciation. The complexity of genome evolution poses many exciting challenges to developers of mathematical models and algorithms, who have recourse to a spectrum of algorithmic, statistical and mathematical techniques, ranging from exact, heuristics, fixed parameter and approximation algorithms for problems based on parsimony models to Markov chain Monte Carlo algorithms for Bayesian analysis of problems based on probabilistic models.

Many of these studies are based on the detection of sequence homology to assign sequences to protein families.

Pan Genomics

Pan genomics is a concept introduced in 2005 by Tettelin and Medini which eventually took root in bioinformatics. Pan genome is the complete gene repertoire of a particular taxonomic group: although initially applied to closely related strains of a species, it can be applied to a larger context like genus, phylum etc. It is divided in two parts- The Core genome: Set of genes common to all the genomes under study (These are often housekeeping genes vital for survival) and The Dispensable/Flexible Genome: Set of genes not present in all but one or some genomes under study. A bioinformatics tool BPGA can be used to characterize the Pan Genome of bacterial species.

Genetics of Disease

With the advent of next-generation sequencing we are obtaining enough sequence data to map the genes of complex diseases infertility, breast cancer or Alzheimer's disease. Genome-wide association studies are a useful approach to pinpoint the mutations responsible for such complex diseases. Through these studies, thousands of DNA variants have been identified that are associated with similar diseases and traits. Furthermore, the possibility for genes to be used at prognosis, diagnosis or treatment is one of the most essential applications. Many studies are discussing both the promising ways to choose the genes to be used and the problems and pitfalls of using genes to predict disease presence or prognosis.

Analysis of Mutations in Cancer

In cancer, the genomes of affected cells are rearranged in complex or even unpredictable ways. Massive sequencing efforts are used to identify previously unknown point mutations in a variety of genes in cancer. Bioinformaticians continue to produce specialized automated systems to manage the sheer volume of sequence data produced, and they create new algorithms and software to compare the sequencing results to the growing collection of human genome sequences and germline polymorphisms. New physical detection technologies are employed, such as oligonucleotide microarrays to identify chromosomal gains and losses (called comparative genomic hybridization), and single-nucleotide polymorphism arrays to detect known *point mutations*. These detection methods simultaneously measure several hundred thousand sites throughout the genome, and when used in high-throughput to measure thousands of samples, generate terabytes of data per experiment. Again the massive amounts and new types of data generate new opportunities for bioinformaticians. The data is often found to contain considerable variability, or noise, and thus Hidden Markov model and change-point analysis methods are being developed to infer real copy number changes.

Two important principles can be used in the analysis of cancer genomes bioinformatically pertaining to the identification of mutations in the exome. First, cancer is a disease of accumulated somatic mutations in genes. Second cancer contains driver mutations which need to be distinguished from passengers.

With the breakthroughs that this next-generation sequencing technology is providing to the field of Bioinformatics, cancer genomics could drastically change. These new methods and software allow bioinformaticians to sequence many cancer genomes quickly and affordably. This could create a more flexible process for classifying types of cancer by analysis of cancer driven mutations in the genome. Furthermore, tracking of patients while the disease progresses may be possible in the future with the sequence of cancer samples.

Another type of data that requires novel informatics development is the analysis of lesions found to be recurrent among many tumors.

Gene and Protein Expression

Analysis of Gene Expression

The expression of many genes can be determined by measuring mRNA levels with multiple techniques including microarrays, expressed cDNA sequence tag (EST) sequencing, serial analysis of gene expression (SAGE) tag sequencing, massively parallel signature sequencing (MPSS), RNA-Seq, also known as "Whole Transcriptome Shotgun Sequencing" (WTSS), or various applications of multiplexed in-situ hybridization. All of these techniques are extremely noise-prone and/or subject to bias in the biological measurement, and a major research area in computational biology involves developing statistical tools to separate signal from noise in high-throughput gene expression studies. Such studies are often used to determine the genes implicated in a disorder: one might compare microarray data from cancerous epithelial cells to data from non-cancerous cells to determine the transcripts that are up-regulated and down-regulated in a particular population of cancer cells.

Analysis of Protein Expression

Protein microarrays and high throughput (HT) mass spectrometry (MS) can provide a snapshot of the proteins present in a biological sample. Bioinformatics is very much involved in making sense of protein microarray and HT MS data; the former approach faces similar problems as with microarrays targeted at mRNA, the latter involves the problem of matching large amounts of mass data against predicted masses from protein sequence databases, and the complicated statistical analysis of samples where multiple, but incomplete peptides from each protein are detected. Cellular protein localization in a tissue context can be achieved through affinity proteomics displayed as spatial data based on immunohistochemistry and tissue microarrays.

Analysis of Regulation

Gene regulation is the complex orchestration of events by which a signal, potentially an extracellular signal such as a hormone, eventually leads to an increase or decrease in the activity of one or more proteins. Bioinformatics techniques have been applied to explore various steps in this process.

For example, gene expression can be regulated by nearby elements in the genome. Promoter analysis involves the identification and study of sequence motifs in the DNA surrounding the coding region of a gene. These motifs influence the extent to which that region is transcribed into mRNA. Enhancer elements far away from the promoter can also regulate gene expression, through three-dimensional looping interactions. These interactions can be determined by bioinformatic analysis of chromosome conformation capture experiments.

Expression data can be used to infer gene regulation: one might compare microarray data from a wide variety of states of an organism to form hypotheses about the genes involved in each state. In a single-cell organism, one might compare stages of the cell cycle, along with various stress conditions (heat shock, starvation, etc.). One can then apply clustering algorithms to that expression data to determine which genes are co-expressed. For example, the upstream regions (promoters) of co-expressed genes can be searched for over-represented regulatory elements. Examples of clustering algorithms applied in gene clustering are k-means clustering, self-organizing maps (SOMs), hierarchical clustering, and consensus clustering methods.

Analysis of Cellular Organization

Several approaches have been developed to analyze the location of organelles, genes, proteins, and other components within cells. This is relevant as the location of these components affects the events within a cell and thus helps us to predict the behavior of biological systems. A gene ontology category, *cellular compartment*, has been devised to capture subcellular localization in many biological databases.

Microscopy and Image Analysis

Microscopic pictures allow us to locate both organelles as well as molecules. It may also help us to distinguish between normal and abnormal cells, e.g. in cancer.

Protein Localization

The localization of proteins helps us to evaluate the role of a protein. For instance, if a protein is found in the nucleus it may be involved in gene regulation or splicing. By contrast, if a protein is found in mitochondria, it may be involved in respiration or other metabolic processes. Protein localization is thus an important component of protein function prediction. There are well developed protein subcellular localization prediction resources available, including protein subcellular location databases, and prediction tools.

Nuclear Organization of Chromatin

Data from high-throughput chromosome conformation capture experiments, such as Hi-C (experiment) and ChIA-PET, can provide information on the spatial proximity of DNA loci. Analysis of these experiments can determine the three-dimensional structure and nuclear organization of chromatin. Bioinformatic challenges in this field include partitioning the genome into domains, such as Topologically Associating Domains (TADs), that are organised together in three-dimensional space.

Structural Bioinformatics

Protein structure prediction is another important application of bioinformatics. The amino acid sequence of a protein, the so-called primary structure, can be easily determined from the sequence on the gene that codes for it. In the vast majority of cases, this primary structure uniquely determines a structure in its native environment. (Of course, there are exceptions, such as the bovine spongiform encephalopathy – a.k.a. Mad Cow Disease – prion.) Knowledge of this structure is vital in understanding the function of the protein. Structural information is usually classified as one of *secondary*, *tertiary* and *quaternary* structure. A viable general solution to such predictions remains an open problem. Most efforts have so far been directed towards heuristics that work most of the time.

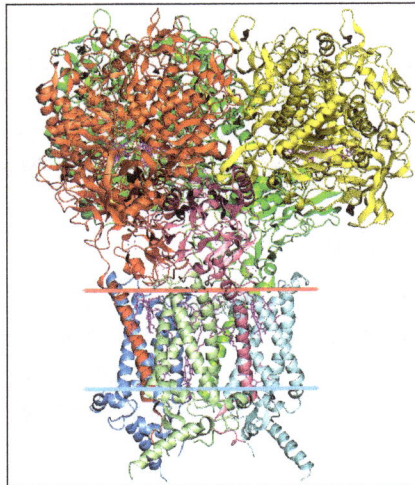

3-dimensional protein structures such as this one are common subjects in bioinformatic analyses.

One of the key ideas in bioinformatics is the notion of homology. In the genomic branch of bioinformatics, homology is used to predict the function of a gene: if the sequence of gene *A*, whose function is known, is homologous to the sequence of gene *B*, whose function is unknown, one could infer that B may share A's function. In the structural branch of bioinformatics, homology is used to determine which parts of a protein are important in structure formation and interaction with other proteins. In a technique called homology modeling, this information is used to predict the structure of a protein once the structure of a homologous protein is known. This currently remains the only way to predict protein structures reliably.

One example of this is hemoglobin in humans and the hemoglobin in legumes (leghemoglobin), which are distant relatives from the same protein superfamily. Both serve the same purpose of transporting oxygen in the organism. Although both of these proteins have completely different amino acid sequences, their protein structures are virtually identical, which reflects their near identical purposes and shared ancestor.

Other techniques for predicting protein structure include protein threading and *de novo* (from scratch) physics-based modeling.

Another aspect of structural bioinformatics include the use of protein structures for Virtual Screening models such as Quantitative Structure-Activity Relationship models and proteochemometric models (PCM). Furthermore, a protein's crystal structure can be used in simulation of for example ligand-binding studies and *in silico* mutagenesis studies.

Network and Systems Biology

Network analysis seeks to understand the relationships within biological networks such as metabolic or protein–protein interaction networks. Although biological networks can be constructed from a single type of molecule or entity (such as genes), network biology often attempts to integrate many different data types, such as proteins, small molecules, gene expression data, and others, which are all connected physically, functionally, or both.

Systems biology involves the use of computer simulations of cellular subsystems (such as the networks of metabolites and enzymes that comprise metabolism, signal transduction pathways and gene regulatory networks) to both analyze and visualize the complex connections of these cellular processes. Artificial life or virtual evolution attempts to understand evolutionary processes via the computer simulation of simple (artificial) life forms.

Molecular Interaction Networks

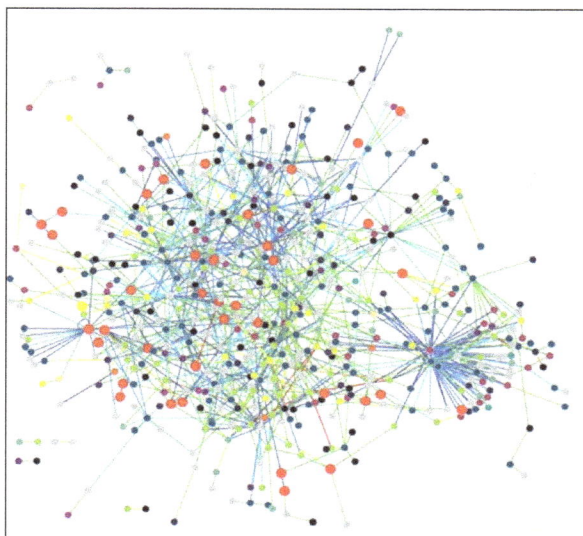

Interactions between proteins are frequently visualized and analyzed using networks. This network is made up of protein–protein interactions from *Treponema pallidum*, the causative agent of syphilis and other diseases.

Tens of thousands of three-dimensional protein structures have been determined by X-ray crystallography and protein nuclear magnetic resonance spectroscopy (protein NMR) and a central question in structural bioinformatics is whether it is practical to predict possible protein–protein interactions only based on these 3D shapes, without performing protein–protein interaction experiments. A variety of methods have been developed to tackle the protein–protein docking problem, though it seems that there is still much work to be done in this field.

Other interactions encountered in the field include Protein–ligand (including drug) and protein–peptide. Molecular dynamic simulation of movement of atoms about rotatable bonds is the fundamental principle behind computational algorithms, termed docking algorithms, for studying molecular interactions.

Others

Literature Analysis

The growth in the number of published literature makes it virtually impossible to read every paper, resulting in disjointed sub-fields of research. Literature analysis aims to employ computational and statistical linguistics to mine this growing library of text resources. For example:

- Abbreviation recognition – identify the long-form and abbreviation of biological terms.

- Named entity recognition – recognizing biological terms such as gene names.

- Protein–protein interaction – identify which proteins interact with which proteins from text.

The area of research draws from statistics and computational linguistics.

High-throughput Image Analysis

Computational technologies are used to accelerate or fully automate the processing, quantification and analysis of large amounts of high-information-content biomedical imagery. Modern image analysis systems augment an observer's ability to make measurements from a large or complex set of images, by improving accuracy, objectivity, or speed. A fully developed analysis system may completely replace the observer. Although these systems are not unique to biomedical imagery, biomedical imaging is becoming more important for both diagnostics and research. Some examples are:

- High-throughput and high-fidelity quantification and sub-cellular localization (high-content screening, cytohistopathology, Bioimage informatics).

- Morphometrics.

- Clinical image analysis and visualization.

- Determining the real-time air-flow patterns in breathing lungs of living animals.

- Quantifying occlusion size in real-time imagery from the development of and recovery during arterial injury.

- Making behavioral observations from extended video recordings of laboratory animals.

- Infrared measurements for metabolic activity determination.

- Inferring clone overlaps in DNA mapping, e.g. the Sulston score.

High-throughput Single Cell Data Analysis

Computational techniques are used to analyse high-throughput, low-measurement single cell data, such as that obtained from flow cytometry. These methods typically involve finding populations of cells that are relevant to a particular disease state or experimental condition.

Biodiversity Informatics

Biodiversity informatics deals with the collection and analysis of biodiversity data, such as taxonomic databases, or microbiome data. Examples of such analyses include phylogenetics, niche modelling, species richness mapping, DNA barcoding, or species identification tools.

Ontologies and Data Integration

Biological ontologies are directed acyclic graphs of controlled vocabularies. They are designed to capture biological concepts and descriptions in a way that can be easily categorised and analysed with computers. When categorised in this way, it is possible to gain added value from holistic and integrated analysis.

The OBO Foundry was an effort to standardise certain ontologies. One of the most widespread is the Gene ontology which describes gene function. There are also ontologies which describe phenotypes.

Databases

Databases are essential for bioinformatics research and applications. Many databases exist, covering various information types: for example, DNA and protein sequences, molecular structures, phenotypes and biodiversity. Databases may contain empirical data (obtained directly from experiments), predicted data (obtained from analysis), or, most commonly, both. They may be specific to a particular organism, pathway or molecule of interest. Alternatively, they can incorporate data compiled from multiple other databases. These databases vary in their format, access mechanism, and whether they are public or not.

Some of the most commonly used databases are listed below:

- Used in biological sequence analysis: Genbank, UniProt.

- Used in structure analysis: Protein Data Bank (PDB).

- Used in finding Protein Families and Motif Finding: InterPro, Pfam.

- Used for Next Generation Sequencing: Sequence Read Archive.

- Used in Network Analysis: Metabolic Pathway Databases (KEGG, BioCyc), Interaction Analysis Databases, Functional Networks.

- Used in design of synthetic genetic circuits: GenoCAD.

Software and Tools

Software tools for bioinformatics range from simple command-line tools, to more complex graphical programs and standalone web-services available from various bioinformatics companies or public institutions.

Open-source Bioinformatics Software

Many free and open-source software tools have existed and continued to grow since the 1980s. The combination of a continued need for new algorithms for the analysis of emerging types of biological readouts, the potential for innovative *in silico* experiments, and freely available open code bases have helped to create opportunities for all research groups to contribute to both bioinformatics and the range of open-source software available, regardless of their funding arrangements. The open source tools often act as incubators of ideas, or community-supported plug-ins in commercial applications. They may also provide *de facto* standards and shared object models for assisting with the challenge of bioinformation integration.

The range of open-source software packages includes titles such as Bioconductor, BioPerl, Biopython, BioJava, BioJS, BioRuby, Bioclipse, EMBOSS, .NET Bio, Orange with its bioinformatics add-on, Apache Taverna, UGENE and GenoCAD. To maintain this tradition and create further opportunities, the non-profit Open Bioinformatics Foundation have supported the annual Bioinformatics Open Source Conference (BOSC) since 2000.

Web Services in Bioinformatics

SOAP- and REST-based interfaces have been developed for a wide variety of bioinformatics applications allowing an application running on one computer in one part of the world to use algorithms, data and computing resources on servers in other parts of the world. The main advantages derive from the fact that end users do not have to deal with software and database maintenance overheads.

Basic bioinformatics services are classified by the EBI into three categories: SSS (Sequence Search Services), MSA (Multiple Sequence Alignment), and BSA (Biological Sequence Analysis). The availability of these service-oriented bioinformatics resources demonstrate the applicability of web-based bioinformatics solutions, and range from a collection of standalone tools with a common data format under a single, standalone or web-based interface, to integrative, distributed and extensible bioinformatics workflow management systems.

Bioinformatics Workflow Management Systems

A bioinformatics workflow management system is a specialized form of a workflow management system designed specifically to compose and execute a series of computational or data manipulation steps, or a workflow, in a Bioinformatics application. Such systems are designed to:

- Provide an easy-to-use environment for individual application scientists themselves to create their own workflows,

- Provide interactive tools for the scientists enabling them to execute their workflows and view their results in real-time,

- Simplify the process of sharing and reusing workflows between the scientists,

- Enable scientists to track the provenance of the workflow execution results and the workflow creation steps.

Some of the platforms giving this service: Galaxy, Kepler, Taverna, UGENE, Anduril, HIVE.

BioCompute and BioCompute Objects

In 2014, the US Food and Drug Administration sponsored a conference held at the National Institutes of Health Bethesda Campus to discuss reproducibility in bioinformatics. Over the next three years, a consortium of stakeholders met regularly to discuss what would become BioCompute paradigm. These stakeholders included representatives from government, industry, and academic entities. Session leaders represented numerous branches of the FDA and NIH Institutes and Centers, non-profit entities including the Human Variome Project and the European Federation for Medical Informatics, and research institutions including Stanford, the New York Genome Center, and the George Washington University.

It was decided that the BioCompute paradigm would be in the form of digital 'lab notebooks' which allow for the reproducibility, replication, review, and reuse, of bioinformatics protocols. This was proposed to enable greater continuity within a research group over the course of normal personnel flux while furthering the exchange of ideas between groups. The US FDA funded this work so that information on pipelines would be more transparent and accessible to their regulatory staff.

Biomedical Engineering

Biomedical engineering (BME) or medical engineering is the application of engineering principles and design concepts to medicine and biology for healthcare purposes (e.g. diagnostic or therapeutic). This field seeks to close the gap between engineering and medicine, combining the design and problem solving skills of engineering with medical biological sciences to advance health care treatment, including diagnosis, monitoring, and therapy. Also included under the scope of a biomedical engineer is the management of current medical equipment within hospitals while adhering to relevant industry standards. This involves equipment recommendations, procurement, routine testing and preventative maintenance, through to decommissioning and disposal. This role is also known as a Biomedical Equipment Technician (BMET) or clinical engineering.

Biomedical engineering has recently emerged as its own study, as compared to many other engineering fields. Such an evolution is common as a new field transition from being an interdisciplinary specialization among already-established fields, to being considered a field in itself. Much of the work in biomedical engineering consists of research and development, spanning a broad array of subfields. Prominent biomedical engineering applications include the development of biocompatible prostheses, various diagnostic and therapeutic medical devices ranging from clinical equipment to micro-implants, common imaging equipment such as MRIs and EKG/ECGs, regenerative tissue growth, pharmaceutical drugs and therapeutic biologicals.

Medical Devices

This is an *extremely broad category*—essentially covering all health care products that do not achieve their intended results through predominantly chemical (e.g., pharmaceuticals) or biological (e.g., vaccines) means, and do not involve metabolism.

A medical device is intended for use in:

- The diagnosis of disease or other conditions,

- In the cure, mitigation, treatment, or prevention of disease.

Some examples include pacemakers, infusion pumps, the heart-lung machine, dialysis machines, artificial organs, implants, artificial limbs, corrective lenses, cochlear implants, ocular prosthetics, facial prosthetics, somato prosthetics, and dental implants.

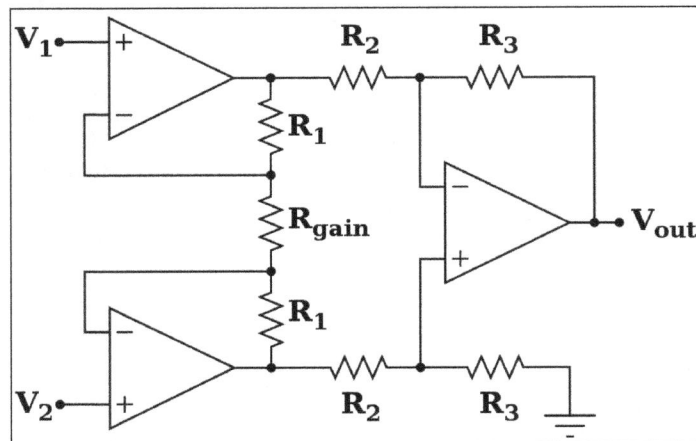

Biomedical instrumentation amplifier schematic used in monitoring low voltage biological signals, an example of a biomedical engineering application of electronic engineering to electrophysiology.

Stereolithography is a practical example of *medical modeling* being used to create physical objects. Beyond modeling organs and the human body, emerging engineering techniques are also currently used in the research and development of new devices for innovative therapies, treatments, patient monitoring, of complex diseases.

Medical devices are regulated and classified (in the US) as follows:

- Class I devices present minimal potential for harm to the user and are often simpler in design than Class II or Class III devices. Devices in this category include tongue depressors, bedpans, elastic bandages, examination gloves, and hand-held surgical instruments and other similar types of common equipment.

- Class II devices are subject to special controls in addition to the general controls of Class I devices. Special controls may include special labeling requirements, mandatory performance standards, and postmarket surveillance. Devices in this class are typically non-invasive and include X-ray machines, PACS, powered wheelchairs, infusion pumps, and surgical drapes.

- Class III devices generally require premarket approval (PMA) or premarket notification (510k), a scientific review to ensure the device's safety and effectiveness, in addition to the general controls of Class I. Examples include replacement heart valves, hip and knee joint implants, silicone gel-filled breast implants, implanted cerebellar stimulators, implantable pacemaker pulse generators and endosseous (intra-bone) implants.

Medical Imaging

Medical/biomedical imaging is a major segment of medical devices. This area deals with enabling clinicians to directly or indirectly "view" things not visible in plain sight (such as due to their size, and/or location). This can involve utilizing ultrasound, magnetism, UV, radiology, and other means.

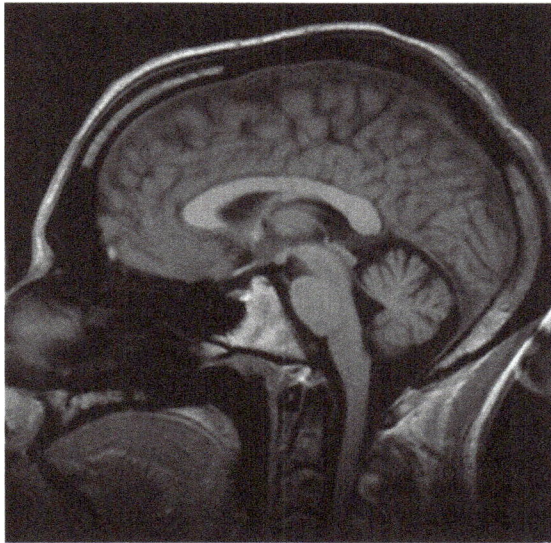

An MRI scan of a human head, an example of a biomedical engineering application of electrical engineering to diagnostic imaging. Click here to view an animated sequence of slices.

Imaging technologies are often essential to medical diagnosis, and are typically the most complex equipment found in a hospital including: fluoroscopy, magnetic resonance imaging (MRI), nuclear medicine, positron emission tomography (PET), PET-CT scans, projection radiography such as X-rays and CT scans, tomography, ultrasound, optical microscopy, and electron microscopy.

Implants

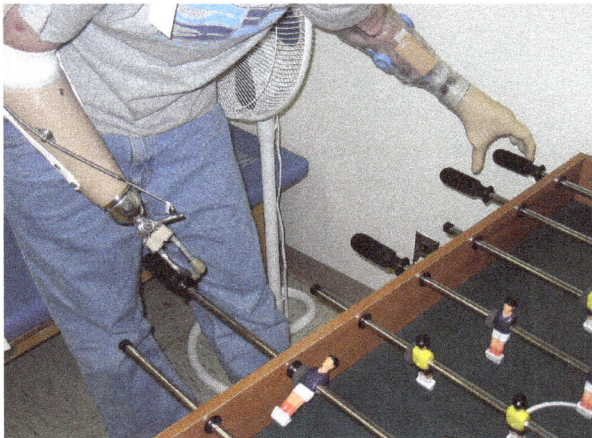

Artificial limbs: The right arm is an example of a prosthesis, and the left arm is an example of myoelectric control.

An implant is a kind of medical device made to replace and act as a missing biological structure (as compared with a transplant, which indicates transplanted biomedical tissue). The surface of implants that contact the body might be made of a biomedical material such as titanium, silicone or apatite depending on what is the most functional. In some cases, implants contain electronics, e.g. artificial pacemakers and cochlear implants. Some implants are bioactive, such as subcutaneous drug delivery devices in the form of implantable pills or drug-eluting stents.

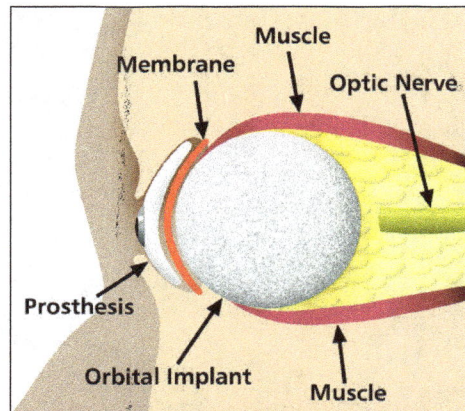

A prosthetic eye, an example of a biomedical engineering application of mechanical engineering and biocompatible materials to ophthalmology.

Bionics

Artificial body part replacements are one of the many applications of bionics. Concerned with the intricate and thorough study of the properties and function of human body systems, bionics may be applied to solve some engineering problems. Careful study of the different functions and processes of the eyes, ears, and other organs paved the way for improved cameras, television, radio transmitters and receivers, and many other useful tools. These developments have indeed made our lives better, but the best contribution that bionics has made is in the field of biomedical engineering (the building of useful replacements for various parts of the human body). Modern hospitals now have available spare parts to replace body parts badly damaged by injury or disease . Biomedical engineers work hand in hand with doctors to build these artificial body parts.

Biomedical Sensors

In recent years biomedical sensors based in microwave technology have gained more attention. Different sensors can be manufactured for specific uses in both diagnosing and monitoring disease conditions, for example microwave sensors can be used as a complementary technique to X-ray to monitor lower extremity trauma. The sensor monitor the dielectric properties and can thus notice change in tissue (bone, muscle, fat etc.) under the skin so when measuring at different times during the healing process the response from the sensor will change as the trauma heals.

Clinical Engineering

Clinical engineering is the branch of biomedical engineering dealing with the actual implementation of medical equipment and technologies in hospitals or other clinical settings. Major roles of clinical engineers include training and supervising biomedical equipment technicians (BMETs), selecting technological products/services and logistically managing their implementation, working with governmental regulators on inspections/audits, and serving as technological consultants for other hospital staff (e.g. physicians, administrators, I.T., etc.). Clinical engineers also advise and collaborate with medical device producers regarding prospective design improvements based on clinical experiences, as well as monitor the progression of the state of the art so as to redirect procurement patterns accordingly.

Their inherent focus on *practical* implementation of technology has tended to keep them oriented more towards *incremental*-level redesigns and reconfigurations, as opposed to revolutionary research & development or ideas that would be many years from clinical adoption; however, there is a growing effort to expand this time-horizon over which clinical engineers can influence the trajectory of biomedical innovation. In their various roles, they form a "bridge" between the primary designers and the end-users, by combining the perspectives of being both 1) close to the point-of-use, while 2) trained in product and process engineering. Clinical engineering departments will sometimes hire not just biomedical engineers, but also industrial/systems engineers to help address operations research/optimization, human factors, cost analysis, etc. There are various companies that help biomedical engineers to bring out their plans into action by making medical devices/machines. Few of the top companies include GE, PHILIPS, SIEMENS, DRAGER, DRAQIL etc.

Rehabilitation Engineering

Rehabilitation engineering is the systematic application of engineering sciences to design, develop, adapt, test, evaluate, apply, and distribute technological solutions to problems confronted by individuals with disabilities. Functional areas addressed through rehabilitation engineering may include mobility, communications, hearing, vision, and cognition, and activities associated with employment, independent living, education, and integration into the community.

While some rehabilitation engineers have master's degrees in rehabilitation engineering, usually a subspecialty of Biomedical engineering, most rehabilitation engineers have undergraduate or graduate degrees in biomedical engineering, mechanical engineering, or electrical engineering. A Portuguese university provides an undergraduate degree and a master's degree in Rehabilitation Engineering and Accessibility. Qualification to become a Rehab' Engineer in the UK is possible via a University BSc Honours Degree course such as Health Design & Technology Institute, Coventry University.

The rehabilitation process for people with disabilities often entails the design of assistive devices such as Walking aids intended to promote inclusion of their users into the mainstream of society, commerce, and recreation.

Molecular Biology

Molecular biology is a branch of biology that concerns the molecular basis of biological activity between biomolecules in the various systems of a cell, including the interactions between DNA, RNA, proteins and their biosynthesis, as well as the regulation of these interactions.

Molecular biology is not so much a technique as an approach an approach from the viewpoint of the so-called basic sciences with the leading idea of searching below the large-scale manifestations of classical biology for the corresponding molecular plan. It is concerned particularly with the *forms* of biological molecules and is predominantly three-dimensional and structural – which does not mean, however, that it is merely a refinement of morphology. It must at the same time inquire into genesis and function.

Relationship to other Biological Sciences

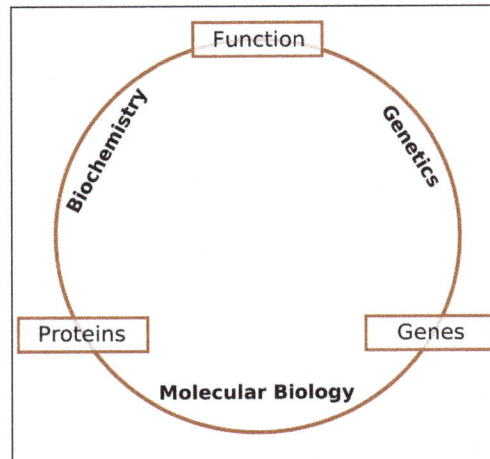

Schematic relationship between biochemistry, genetics and molecular biology.

Researchers in molecular biology use specific techniques native to molecular biology but increasingly combine these with techniques and ideas from genetics and biochemistry. There is not a defined line between these disciplines. This is shown in the following schematic that depicts one possible view of the relationships between the fields:

- Biochemistry is the study of the chemical substances and vital processes occurring in live organisms. Biochemists focus heavily on the role, function, and structure of biomolecules. The study of the chemistry behind biological processes and the synthesis of biologically active molecules are examples of biochemistry.

- Genetics is the study of the effect of genetic differences in organisms. This can often be inferred by the absence of a normal component (e.g. one gene). The study of "mutants" – organisms which lack one or more functional components with respect to the so-called "wild type" or normal phenotype. Genetic interactions (epistasis) can often confound simple interpretations of such "knockout" studies.

- Molecular biology is the study of molecular underpinnings of the processes of replication, transcription, translation, and cell function. The central dogma of molecular biology where genetic material is transcribed into RNA and then translated into protein, despite being oversimplified, still provides a good starting point for understanding the field. The picture has been revised in light of emerging novel roles for RNA.

Much of molecular biology is quantitative, and recently much work has been done at its interface with computer science in bioinformatics and computational biology. In the early 2000s, the study of gene structure and function, molecular genetics, has been among the most prominent sub-fields of molecular biology. Increasingly many other areas of biology intersecting with molecular Biology, focus on molecules, either directly studying interactions in their own right such as in cell biology and developmental biology, or indirectly, where molecular techniques are used to infer historical attributes of populations or species, as in fields in evolutionary biology such as population genetics and phylogenetics. There is also a long tradition of studying biomolecules "from the ground up" in biophysics.

Techniques of Molecular Biology

Molecular Cloning

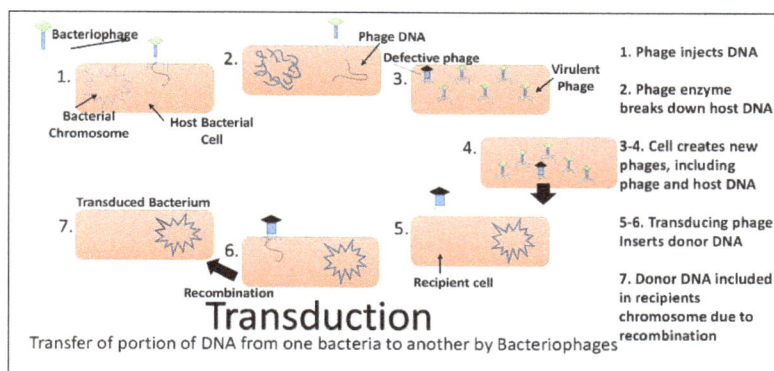

Transduction image.

One of the most basic techniques of molecular biology to study protein function is molecular cloning. In this technique, DNA coding for a protein of interest is cloned using polymerase chain reaction (PCR), and/or restriction enzymes into a plasmid (expression vector). A vector has 3 distinctive features: an origin of replication, a multiple cloning site (MCS), and a selective marker usually antibiotic resistance. Located upstream of the multiple cloning site are the promoter regions and the transcription start site which regulate the expression of cloned gene. This plasmid can be inserted into either bacterial or animal cells. Introducing DNA into bacterial cells can be done by transformation via uptake of naked DNA, conjugation via cell-cell contact or by transduction via viral vector. Introducing DNA into eukaryotic cells, such as animal cells, by physical or chemical means is called transfection. Several different transfection techniques are available, such as calcium phosphate transfection, electroporation, microinjection and liposome transfection. The plasmid may be integrated into the genome, resulting in a stable transfection, or may remain independent of the genome, called transient transfection.

DNA coding for a protein of interest is now inside a cell, and the protein can now be expressed. A variety of systems, such as inducible promoters and specific cell-signaling factors, are available to help express the protein of interest at high levels. Large quantities of a protein can then be extracted from the bacterial or eukaryotic cell. The protein can be tested for enzymatic activity under a variety of situations, the protein may be crystallized so its tertiary structure can be studied, or, in the pharmaceutical industry, the activity of new drugs against the protein can be studied.

Polymerase Chain Reaction

Polymerase chain reaction (PCR) is an extremely versatile technique for copying DNA. In brief, PCR allows a specific DNA sequence to be copied or modified in predetermined ways. The reaction is extremely powerful and under perfect conditions could amplify one DNA molecule to become 1.07 billion molecules in less than two hours. The PCR technique can be used to introduce restriction enzyme sites to ends of DNA molecules, or to mutate particular bases of DNA, the latter is a method referred to as site-directed mutagenesis. PCR can also be used to determine whether a particular DNA fragment is found in a cDNA library. PCR has many variations, like reverse transcription PCR (RT-PCR) for amplification of RNA, and, more recently, quantitative PCR which allow for quantitative measurement of DNA or RNA molecules.

Gel Electrophoresis

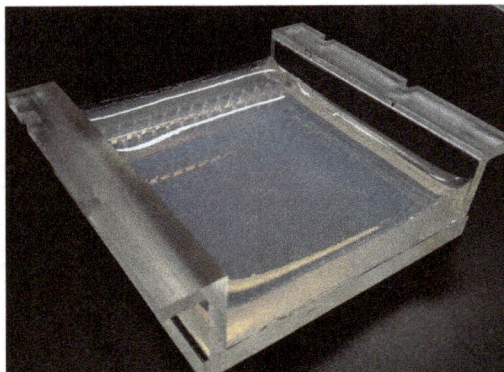

Two percent agarose gel in borate buffer cast in a gel tray.

Gel electrophoresis is one of the principal tools of molecular biology. The basic principle is that DNA, RNA, and proteins can all be separated by means of an electric field and size. In agarose gel electrophoresis, DNA and RNA can be separated on the basis of size by running the DNA through an electrically charged agarose gel. Proteins can be separated on the basis of size by using an SDS-PAGE gel, or on the basis of size and their electric charge by using what is known as a 2D gel electrophoresis.

Macromolecule Blotting and Probing

The terms *northern*, *western* and *eastern* blotting are derived from what initially was a molecular biology joke that played on the term *Southern blotting*, after the technique described by Edwin Southern for the hybridisation of blotted DNA. Patricia Thomas, developer of the RNA blot which then became known as the *northern blot*, actually didn't use the term.

Southern Blotting

Named after its inventor, biologist Edwin Southern, the Southern blot is a method for probing for the presence of a specific DNA sequence within a DNA sample. DNA samples before or after restriction enzyme (restriction endonuclease) digestion are separated by gel electrophoresis and then transferred to a membrane by blotting via capillary action. The membrane is then exposed to a labeled DNA probe that has a complement base sequence to the sequence on the DNA of interest. Southern blotting is less commonly used in laboratory science due to the capacity of other techniques, such as PCR, to detect specific DNA sequences from DNA samples. These blots are still used for some applications, however, such as measuring transgene copy number in transgenic mice or in the engineering of gene knockout embryonic stem cell lines.

Northern Blotting

The northern blot is used to study the expression patterns of a specific type of RNA molecule as relative comparison among a set of different samples of RNA. It is essentially a combination of denaturing RNA gel electrophoresis, and a blot. In this process RNA is separated based on size and is then transferred to a membrane that is then probed with a labeled complement of a sequence of interest. The results may be visualized through a variety of ways depending on the label used; however, most result in the revelation of bands representing the sizes of the RNA detected in sample. The intensity of these bands is related to the amount of the target RNA in the samples analyzed. The procedure is commonly used to study when and how much gene expression is occurring by measuring how much of that RNA is present in different samples. It is one of the most basic tools for determining at what time, and under what conditions, certain genes are expressed in living tissues.

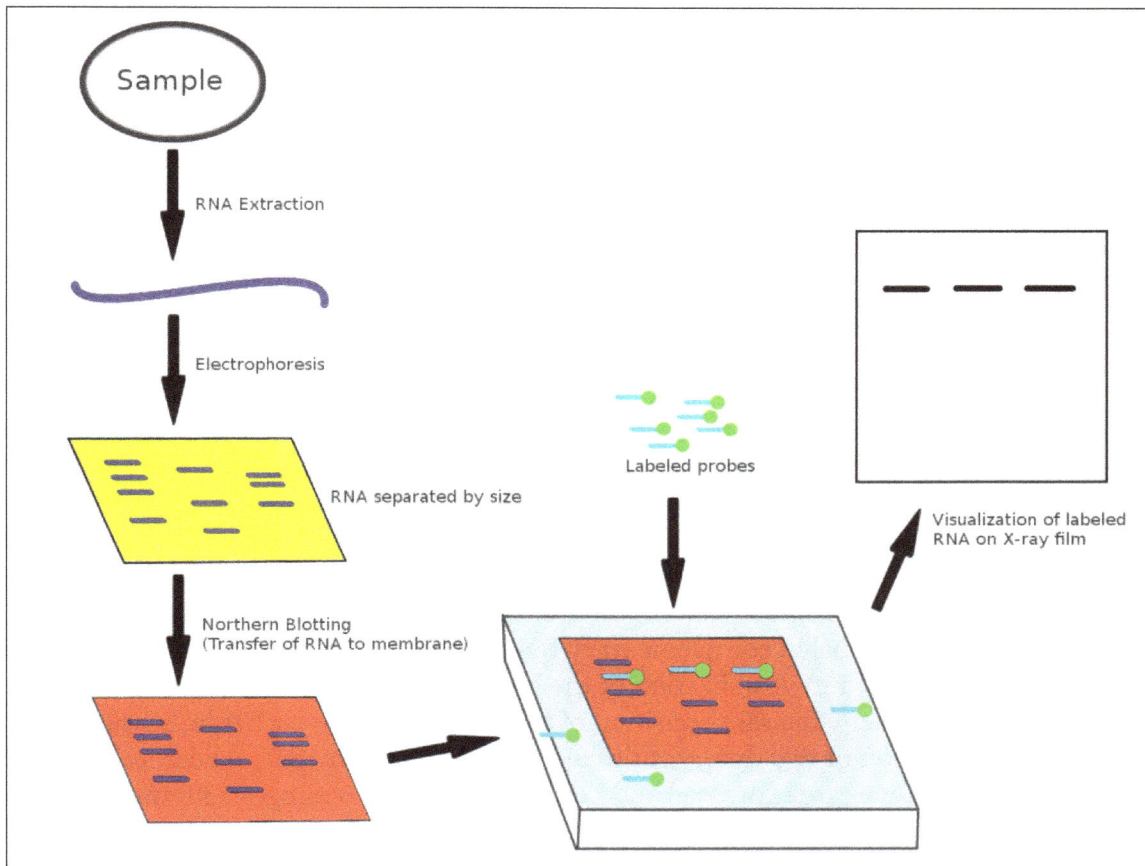

Northern blot diagram.

Western Blotting

In western blotting, proteins are first separated by size, in a thin gel sandwiched between two glass plates in a technique known as SDS-PAGE. The proteins in the gel are then transferred to a polyvinylidene fluoride (PVDF), nitrocellulose, nylon, or other support membrane. This membrane can then be probed with solutions of antibodies. Antibodies that specifically bind to the protein of interest can then be visualized by a variety of techniques, including colored products, chemiluminescence, or autoradiography. Often, the antibodies are labeled with enzymes. When a chemiluminescent substrate is exposed to the enzyme it allows detection. Using western blotting techniques allows not only detection but also quantitative analysis. Analogous methods to western blotting can be used to directly stain specific proteins in live cells or tissue sections.

Eastern Blotting

The eastern blotting technique is used to detect post-translational modification of proteins. Proteins blotted on to the PVDF or nitrocellulose membrane are probed for modifications using specific substrates.

Microarrays

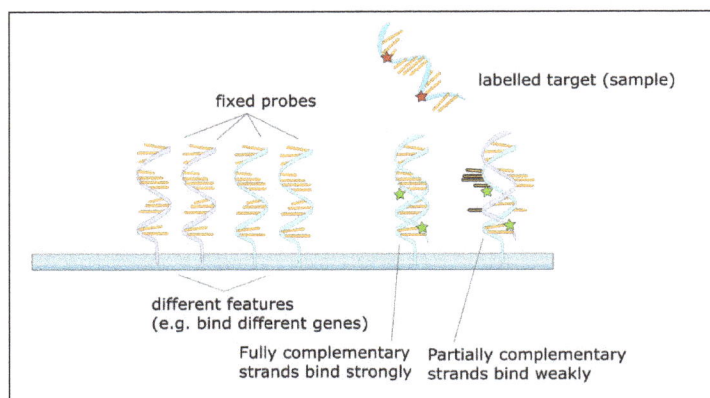

Hybridization of target to probe.

A DNA microarray is a collection of spots attached to a solid support such as a microscope slide where each spot contains one or more single-stranded DNA oligonucleotide fragments. Arrays make it possible to put down large quantities of very small (100 micrometre diameter) spots on a single slide. Each spot has a DNA fragment molecule that is complementary to a single DNA sequence. A variation of this technique allows the gene expression of an organism at a particular stage in development to be qualified (expression profiling). In this technique the RNA in a tissue is isolated and converted to labeled complementary DNA (cDNA). This cDNA is then hybridized to the fragments on the array and visualization of the hybridization can be done. Since multiple arrays can be made with exactly the same position of fragments they are particularly useful for comparing the gene expression of two different tissues, such as a healthy and cancerous tissue. Also, one can measure what genes are expressed and how that expression changes with time or with other factors. There are many different ways to fabricate microarrays; the most common are silicon chips, microscope slides with spots of ~100 micrometre diameter, custom arrays, and arrays with larger spots on porous membranes (macroarrays). There can be anywhere from 100 spots to more than 10,000 on a given array. Arrays can also be made with molecules other than DNA.

Allele-specific Oligonucleotide

Allele-specific oligonucleotide (ASO) is a technique that allows detection of single base mutations without the need for PCR or gel electrophoresis. Short (20–25 nucleotides in length), labeled probes are exposed to the non-fragmented target DNA, hybridization occurs with high specificity due to the short length of the probes and even a single base change will hinder hybridization. The target DNA is then washed and the labeled probes that didn't hybridize are removed. The target DNA is then analyzed for the presence of the probe via radioactivity or fluorescence. In this experiment, as in most molecular biology techniques, a control must be used to ensure successful experimentation.

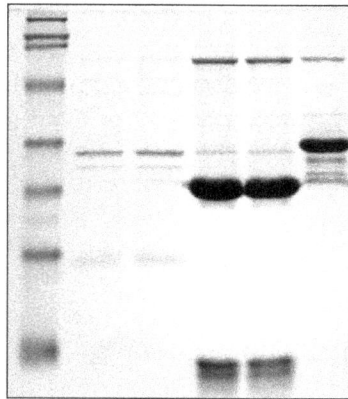
SDS-PAGE.

In molecular biology, procedures and technologies are continually being developed and older technologies abandoned. For example, before the advent of DNA gel electrophoresis (agarose or polyacrylamide), the size of DNA molecules was typically determined by rate sedimentation in sucrose gradients, a slow and labor-intensive technique requiring expensive instrumentation; prior to sucrose gradients, viscometry was used. Aside from their historical interest, it is often worth knowing about older technology, as it is occasionally useful to solve another new problem for which the newer technique is inappropriate.

Biomedicine

Biomedicine (i.e. medical biology) is a branch of medical science that applies biological and physiological principles to clinical practice. The branch especially applies to biology and physiology. Biomedicine also can relate to many other categories in health and biological related fields. It has been the dominant health system for more than a century.

It includes many biomedical disciplines and areas of specialty that typically contain the "bio-" prefix such as molecular biology, biochemistry, biotechnology, cell biology, embryology, nanobiotechnology, biological engineering, laboratory medical biology, cytogenetics, genetics, gene therapy, bioinformatics, biostatistics, systems biology, neuroscience, microbiology, virology, immunology, parasitology, physiology, pathology, anatomy, toxicology, and many others that generally concern life sciences as applied to medicine.

Medical biology is the cornerstone of modern health care and laboratory diagnostics. It concerns a wide range of scientific and technological approaches: from in vitro diagnostics to in vitro fertilisation, from the molecular mechanisms of cystic fibrosis to the population dynamics of the HIV virus, from the understanding of molecular interactions to the study of carcinogenesis, from a single-nucleotide polymorphism (SNP) to gene therapy.

Medical biology is based on molecular biology and combines all issues of developing molecular medicine into large-scale structural and functional relationships of the human genome, transcriptome, proteome, physiome and metabolome with the particular point of view of devising new technologies for prediction, diagnosis and therapy.

Biomedicine involves the study of (patho-) physiological processes with methods from biology and physiology. Approaches range from understanding molecular interactions to the study of the consequences at the in vivo level. These processes are studied with the particular point of view of devising new strategies for diagnosis and therapy.

Depending on the severity of the disease, biomedicine pinpoints a problem within a patient and fixes the problem through medical intervention. Medicine focuses on curing diseases rather than improving one's health.

In social sciences biomedicine is described somewhat differently. Through an anthropological lens biomedicine extends beyond the realm of biology and scientific facts; it is a socio-cultural system which collectively represents reality. While biomedicine is traditionally thought to have no bias due to the evidence-based practices, Gaines & Davis-Floyd highlight that biomedicine itself has a cultural basis and this is because biomedicine reflects the norms and values of its creators. Increasingly other cultures are informing broader medical knowledge - and by extension - biomedicine. Another basis for defining biomedicine as a cultural system is the involvement of hierarchies - a common theme within cultures - between health care practitioners and the prioritizing of different groups, such as patient versus doctor, general physician versus surgeon, or young versus old. Gaines and Davis-Floyd's final point of labelling biomedicine as its own culture suggests that studies constructed within the biomedical field tend to aim at confirming previously established knowledge rather than questioning it.

References

- Elsevier. "book series: principles of medical biology". Www.elsevier.com. Archived from the original on 2012-10-10. Retrieved 2013-03-20

- Environmental-biotechnology-meaning-applications-and-other-details, environmental-biotechnology, biotechnology: biologydiscussion.com, Retrieved 19 April, 2019

- "International service for the acquisition of agri-biotech applications - isaaa.org". Www.isaaa.org. Retrieved 2016-12-05

- Nanobiotechnology: nanowerk.com, Retrieved 26 July, 2019

- Sim, a. Y. L.; minary, p.; levitt, m. (2012). "modeling nucleic acids". Current opinion in structural biology. 22 (3): 273–278. Doi:10.1016/j.sbi.2012.03.012. Pmc 4028509. Pmid 22538125

- What-industrial-biotechnology: bio.org, Retrieved 25 February, 2019

- John denis enderle; joseph d. Bronzino (2012). Introduction to biomedical engineering. Academic press. Pp. 16–. Isbn 978-0-12-374979-6

- R. Mcneill alexander (2005) mechanics of animal movement, current biologyvolume 15, issue 16, 23 august 2005, pages r616-r619. Doi:10.1016/j.cub.2005.08.016

- "Biomedicine (applies biological, physiological) – memidex dictionary/thesaurus". Memidex.com. 2012-10-08. Retrieved 2012-10-20

Permissions

All chapters in this book are published with permission under the Creative Commons Attribution Share Alike License or equivalent. Every chapter published in this book has been scrutinized by our experts. Their significance has been extensively debated. The topics covered herein carry significant information for a comprehensive understanding. They may even be implemented as practical applications or may be referred to as a beginning point for further studies.

We would like to thank the editorial team for lending their expertise to make the book truly unique. They have played a crucial role in the development of this book. Without their invaluable contributions this book wouldn't have been possible. They have made vital efforts to compile up to date information on the varied aspects of this subject to make this book a valuable addition to the collection of many professionals and students.

This book was conceptualized with the vision of imparting up-to-date and integrated information in this field. To ensure the same, a matchless editorial board was set up. Every individual on the board went through rigorous rounds of assessment to prove their worth. After which they invested a large part of their time researching and compiling the most relevant data for our readers.

The editorial board has been involved in producing this book since its inception. They have spent rigorous hours researching and exploring the diverse topics which have resulted in the successful publishing of this book. They have passed on their knowledge of decades through this book. To expedite this challenging task, the publisher supported the team at every step. A small team of assistant editors was also appointed to further simplify the editing procedure and attain best results for the readers.

Apart from the editorial board, the designing team has also invested a significant amount of their time in understanding the subject and creating the most relevant covers. They scrutinized every image to scout for the most suitable representation of the subject and create an appropriate cover for the book.

The publishing team has been an ardent support to the editorial, designing and production team. Their endless efforts to recruit the best for this project, has resulted in the accomplishment of this book. They are a veteran in the field of academics and their pool of knowledge is as vast as their experience in printing. Their expertise and guidance has proved useful at every step. Their uncompromising quality standards have made this book an exceptional effort. Their encouragement from time to time has been an inspiration for everyone.

The publisher and the editorial board hope that this book will prove to be a valuable piece of knowledge for students, practitioners and scholars across the globe.

Index